Serial port and microcontrollers

Principles, circuits, and source codes

Grzegorz Niemirowski

Serial port and microcontrollers.
Principles, circuits, and source codes.

Copyright © 2013 by Grzegorz Niemirowski

First edition

Printed in Charleston, SC, United States

ISBN-13: 978-1481908979

Contents

Introduction

It may seem strange to write another book about serial port. It's an old standard created in 1960's, now seen as obsolete by many. Most of modern computers are not equipped with RS-232 ports. But it's naïve to think that serial port is dead. RS-232 and RS-485 standards are widely used in industry and it's not going to change soon. Especially RS-485 thanks to its reliability found many applications in industrial solutions. The main reason for popularity of these standards is a big simplicity of RS-232 and RS-485. Their transmission protocols are very easy to understand and implement, both in hardware and software. This also means that adding serial communication capability to a device generates relatively low cost. RS-232 and RS-485 have more advantages, for example transmission line can be much longer than for USB. Also hobbyists will find RS-232 much easier to play with than with USB. Thus learning about RS-232 will turn out to be good idea and will give valuable knowledge and lots of fun. RS485 may be seen as more serious but is also very interesting and easy to learn and use.

Serial port and microcontrollers

This book is intended to be not only a description of communication standards but also a good source of practical ideas. I think that most fun comes from playing with self-made devices. To make them useful, simple, and flexible we will use modern microcontrollers. This book covers two families: AVR and ARM Cortex-M3. The first one is very popular and widely used by hobbyists. The second one is quite new but gains popularity very rapidly thanks to rich functionality. The book is a guide into basics of serial communication. It was written for IT people who have good knowledge about computers but didn't have time or opportunity to deal with hardware but would like to design simple electronic devices capable of communicate with a computer. On the other hand this book is for electronics engineers and hobbyists who like soldering but don't have much experience with software. Thanks to variety of simple examples everyone interested in playing with custom made devices will find valuable knowledge and inspiration.

Most of hardware tools, including development boards and programmers, presented in following chapters are available at Kamami: www.kamami.com. You can as well use tools provided by other manufacturers or assembled by yourself.

When I was writing this book, I put much effort into make it comprehensible, easy to read and practical. I you find any mistakes or something is unclear, please let me know. If you have any problems with designing own circuits or developing software using materials from this book, let me know as well. Just drop me an e-mail: grzegorz@grzegorz.net

Schematics, source codes, and other files related to this book can be found at http://www.grzegorz.net/rs232/

1. The RS-232 standard

Examples presented in this book show how to easily use RS-232 standard in electronic circuits and give the basic knowledge about RS-232 which is enough for practical use. Anyway the standard itself is not so simple in some details. It was created when many things in technology looked different than now and it has a few features which are obsolete or unimportant nowadays. Therefore it is good to learn RS-232 in details and understand its philosophy.

1.1. History and assumptions

The RS-232 standard (Recommended Standard 232) was created in 1962 by companies wanting to have universal interface for serial communication between DTE (Data Terminal Equipment) devices and a DCE (Data Circuit-terminating Equipment) devices. Yes, I know that this definition doesn't clarify anything. So let's decipher these abbreviations. DTE is a terminal device in communication channel, an endpoint which is a data source or data sink. An example of DTE is a computer where user enters some data which are later transmitted over some channel. A DTE is also a computer which receives these data. So what is DCE? As you may guess, it's a device sitting between DTEs. DCE is connected from one side to DTE and from the other to communication line. Its task is signal conversion and coding. An example of DCE can be a modem connected to telephone line. In the beginning teletypewriters worked as DTE devices. These were a kind of electromechanical typewriters used to enter data to mainframes or send through some communication channel. But it was long time ago. Since then RS-232 was gaining more and more popularity and it was used to connect great variety of devices. The division to DTEs and DCEs started vanishing and is not much important now.

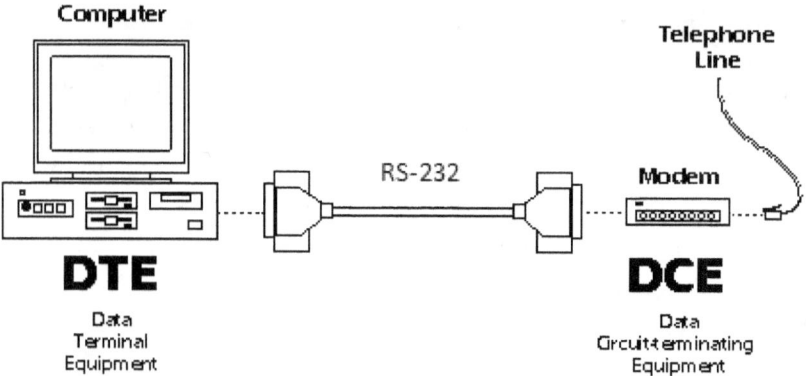

Fig. 1.1.1. Traditional use of RS-232 line as a connection between DTE and DCE

1.2.   Communication lines

To make a communication between DTE and DCE possible, eight lines (wires) are used. The ninth wire is common ground (GND, also known as SG – signal ground); all voltages in other lines are set and measured in reference to it. There are two data lines, TxD (also known as TX or TD) for sending data from DTE to DCE and RxD (also known as RX or RD) for receiving data from DCE to DTE. The rest of lines are control lines. DTR (Data Terminal Ready) is set by DTE and informs DCE about readiness to work (e.g. it was powered on and initialized). Similar line is DSR (Data Set Ready) which tells DTE that DCE is ready. Two signals are used for handshaking: RTS (Request To Send) and CTS (Clear To Send). RTS is set by DTE and tells DCE that it should prepare to receive data from DTE. When DCE is ready for data transmission it sets CTS line which acknowledges former request. The next two lines are typical for modems. DCD (Data Carrier Detect) is set by DCE when it is connected to telephone line. RI (Ring Indicator) line is set by DCE when it detects ring on the telephone line. Nowadays usually only RxD and TxD wires are used which with GND gives three wires. In examples presented in this book such a simple connection is begin used. Sometimes one can take even simpler approach and use only one data line (RxD or TxD) and GND. It is suitable when data are transmitted in only one direction.

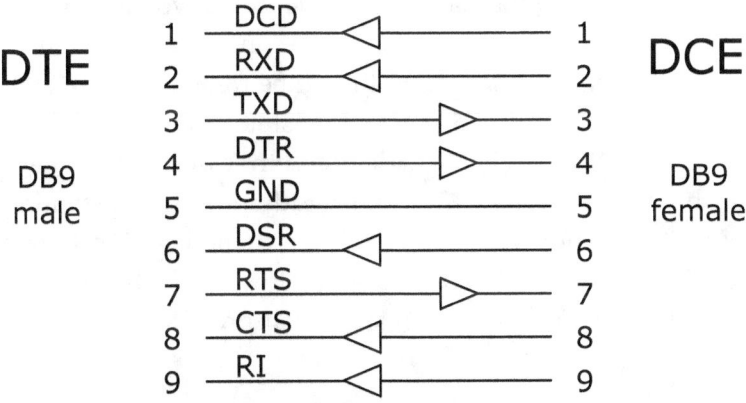

Fig. 1.2.1. RS-232 lines and their directions. The numbers are pin numbers of DB9 connectors, typical for RS-232 cables.

1.3. Voltage levels

Long time ago, when RS-232 standard was created, much higher voltages were used in computer equipment than now. Today two most common voltages in digital systems are 5 V and 3.3 V. Back in 1960's voltage levels could be as high as ±15 volts. Yes, they could be negative with respect to common ground (GND) level. It also means that voltage differences could be as high as 30 volts. RS-232 follows these assumptions. The standard describes high state on the output of a transmitter as a voltage between 5V and 15V. Thus if we want to set a data line to high state, we have to supply a voltage not lower than 5 V and not higher than 15V. On the other hand standard says that the circuit has to withstand a voltage as high as 25 V. What if we want to set a line to low state? We have to power the line from source having voltage between -15 V and -5 V. Thus the voltage has to be at least 5 volts lower than GND and not more than 15 volts lower than GND. Maximum negative voltage which shouldn't break a device is -25 V.

Now one can ask what happens if we set a voltage to a value between -5 and 5 V? How would a receiver interpret this value? If the voltage is between +3 V and +5 V it must be treated as high state and in the area between -5 V and -3 V as low state. So there is 2 V margin of tolerance at the receiver's side in both cases. This interval of voltages between -3 V and +3 V is forbidden in RS-232 standard but usually voltages higher than 1.25 V are treated as high state and voltages lower than 1.0 V are treated as low state. Between 1.0 and 1.25 volts there is area of hysteresis which gives some immunity to noise.

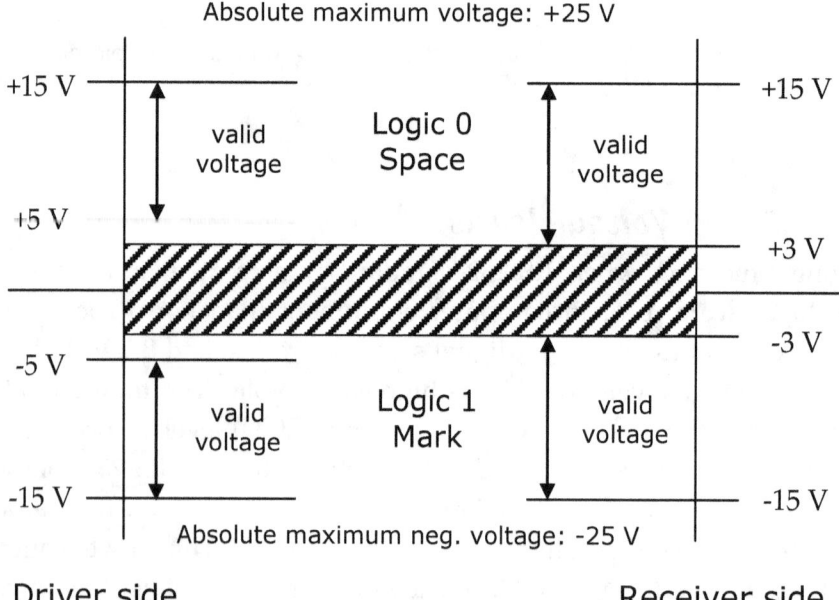

Fig. 1.3.1. RS-232 voltage levels. Valid voltages on transmiter's output – on the left, voltages which should be correctly interpreted by the receiver – on the right.

Knowing this we can go further and ask another question: can't we just connect output from standard digital circuit, powered from 3.3 or 5 volt supply and have everything working? In such case logical zero means 0 V voltage which is below threshold and logical one means 3.3 or 5 V which is above the threshold. Would it work? No and there is more than one reason. First of all it's a nonstandard approach and should be avoided. Secondly there is also problem with incoming signals: ±15 V voltage can damage our device. Last but not least there is a very significant detail: logical states in RS-232 are negated. What does it mean in practice? When logical one is transmitted then the line is in low state, when logical zero is transmitted than the line is in high state. Sometimes you may encounter names Mark and Space. Mark is logical one (low voltage level) and Space is logical zero (high voltage level).

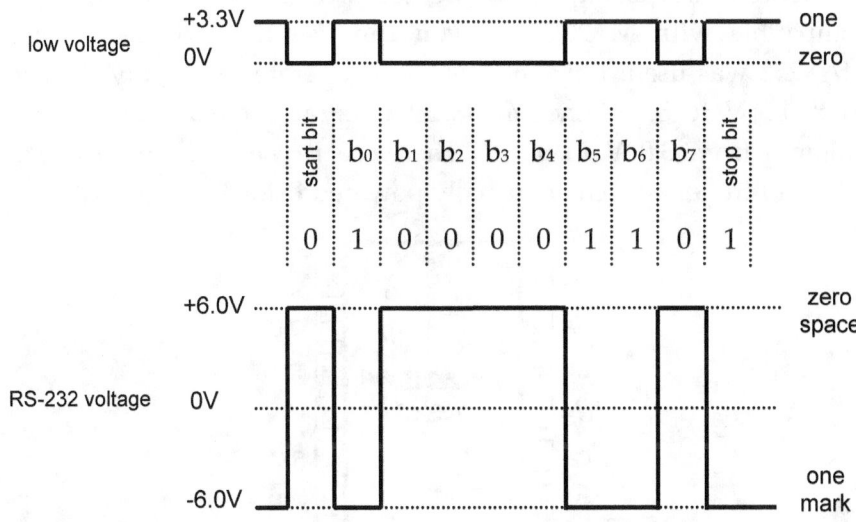

Fig. 1.3.2. Example of voltages in a digital circuit powered from 3.3 V supply compared to an RS-232 line the circuit works with. A byte of value 97 (binary 01100001) is being transmitted.

Problems mentioned above can be solved using simple transistor circuit but there is a better solution: integrated voltage converter chip. You can connect it between a microcontroller and RS-232-

compatible device, e.g. a PC. It is powered from low voltage supply (usually 3 to 5 V) and thanks to built-in capacitive voltage generator it can produce voltages required by RS-232. It can also receive such voltages and provide voltages suitable for a microcontroller. Integrated voltage converter contains not only drivers but also inverters so we don't have to care about logical states being negated on RS-232 line.

The most popular voltage converter IC is MAX232 produced by MAXIM. You can find it in vast majority of devices where voltage conversion for RS-232 is required. MAX232 is quite an old chip however and you should look for more modern ICs. They often have such benefits like higher speeds of transmission, can work with wider power supply range, and require smaller capacitances. Therefore in our examples we are going to use MAX3232. It is pin-compatible with MAX232 so you can use it everywhere where MAX232 was used. But while MAX232 operates at supply voltages from 4.5 V to 5.5 V the MAX3232 integrated circuit can work at voltages from 3.0 V to 5.5 V. Therefore it can work with STM32 microcontrollers which are usually powered from 3.3 V supply.

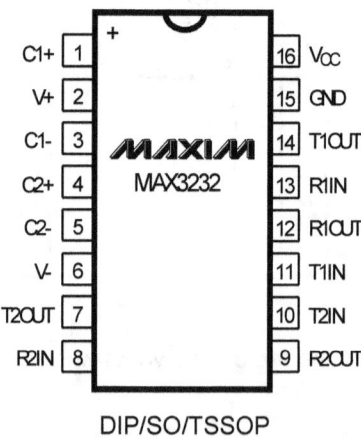

DIP/SO/TSSOP

Fig. 1.3.3. MAX3232 pinout. It's the same for all three packages.

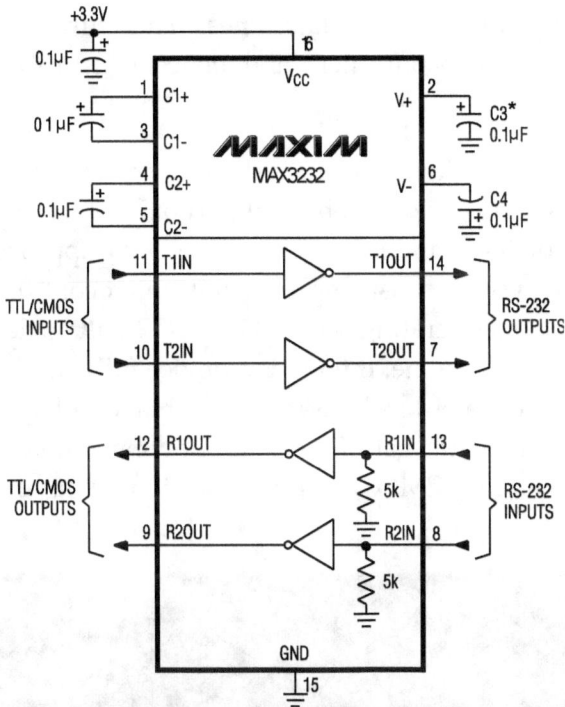

Fig. 1.3.4. Typical operating circuit for MAX3232.

There is one situation when we can forget about converting voltages. Imagine two microcontrollers connected with RS-232 line. Each one of them is accompanied by MAX3232 which does voltage conversion required by RS-232 standard. It's very easy to notice that such solution is overcomplicated. We can ditch voltage conversion and connect microcontrollers directly, connecting TxD pin of the first one to RxD pin of the second one and RxD pin of the first one to TxD pin of the second one because we treat them as two DTEs. Of course we have to connect grounds and have both microcontrollers powered using the same voltage. If the microcontrollers are not powered from the same supply then we need serial resistors on data lines (220 – 470 Ω). Even when provided supply voltages are theoretically the same, like 5V and 5V, in practice there are always some differences which

Serial port and microcontrollers

can cause current flow on data lines. This current can lead to malfunction of microcontrollers and hence resistors are necessary protection.

When we connect microcontrollers directly or with just small value protecting resistors, we drop physical part of RS-232 and use only the logical level of the standard. We can use such approach when we design larger system consisting of a few microcontrollers or other integrated circuits operating with TTL/CMOS voltage levels and RS-232 data encoding scheme. If there is a device fully implementing RS-232, including voltage levels, we have to make voltage conversion. Following pictures are real life signals showing differences between low voltage (TTL/CMOS) and traditional RS-232 voltages.

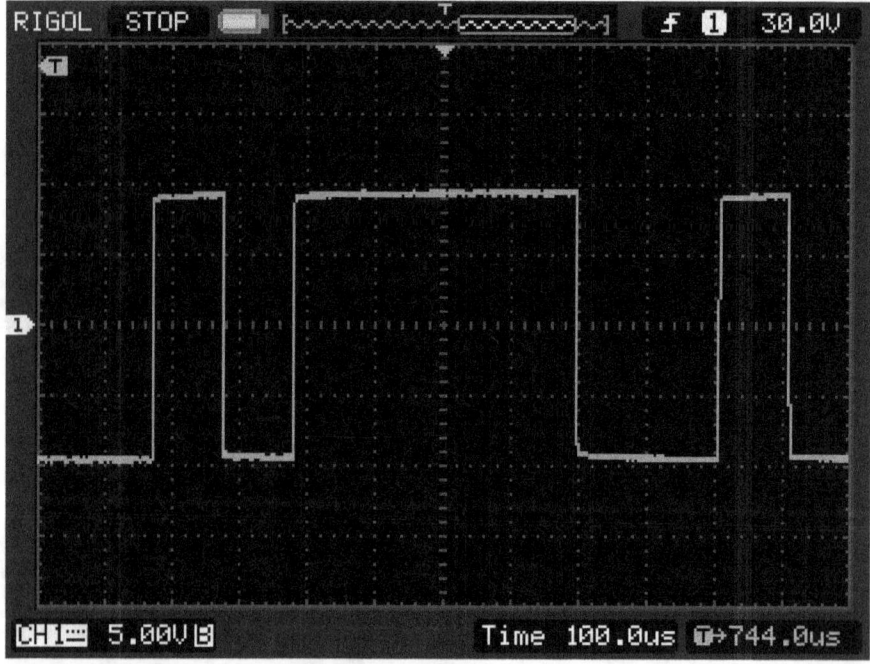

Fig. 1.3.5. RS-232 transmission of letter „a" (byte of value 97, binary 01100001) at 9600 bits per second.

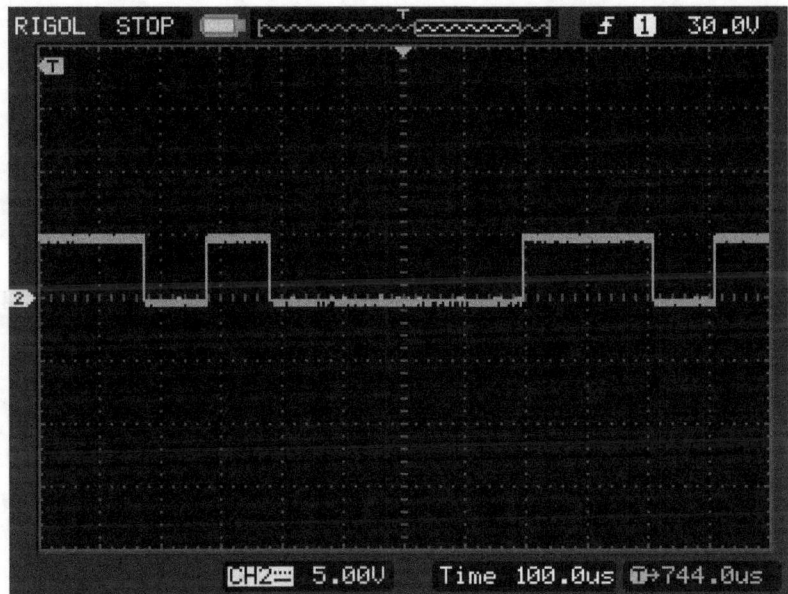

Fig. 1.3.6. The same transmission after voltage conversion.

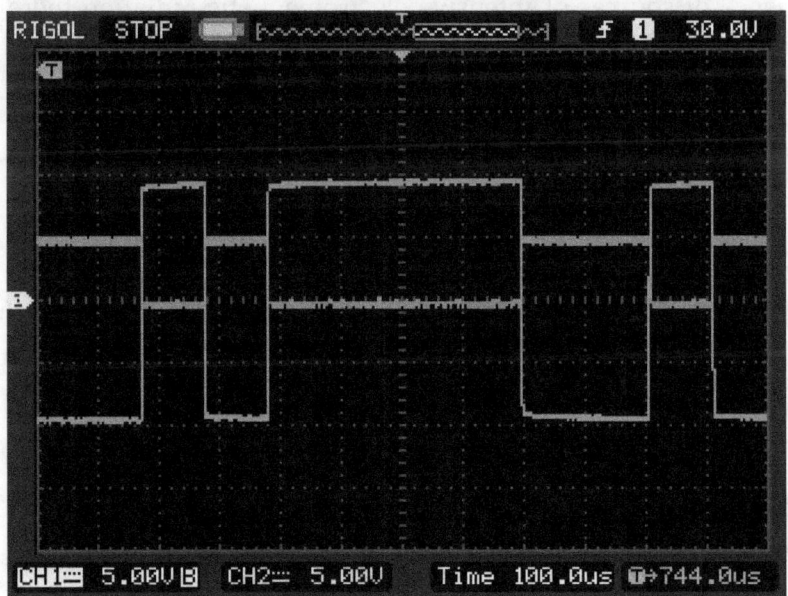

Fig. 1.3.7. Both signals combined.

Serial port and microcontrollers

Pictures 1.3.5, 1.3.6 and 1.3.7 were acquired using an oscilloscope. The first one is original RS-232 signal. The second one was obtained at the output of voltage converter. The last picture shows both signals together.

As we can read, the sensitivity of the oscilloscope is set to 5V per division. Therefore we can see that the voltage levels of RS-232 are about ±10 V here, which is an expected value as valid RS-232 voltage is from ±5 V to ±15 V. The low level voltage is about 5 V which is a typical TTL level. We can also estimate speed of transmission. The time base is set to 100 μs per division. From the picture we can see that width of one bit is a little bit more than one division. Therefore the speed is a little bit lower than 10 000 bps. We can guess that it is 9600 bps – most common bitrate for RS-232.

The next thing we see is noise. Here it's not big and doesn't affect transmission. However with long cables, fast transmissions and in the presence of sources of intensive electromagnetic radiation there can be a problem. Remember that RS-232 wasn't designed for such circumstances and if you require reliability in unfriendly environment you should try RS 485 or other standard designed for immunity for noise.

The other thing we need to consider are rising and falling times of the slopes of signals. Again on presented pictures there is no issue with that. Therefore let's look at another picture (1.3.8). As you can see slopes here don't rise and fall vertically as on previous pictures. It takes about microsecond to achieve valid RS-232 voltage (rise from -10 V to +5 or fall from +10 V to -5V). And it takes about 2 microseconds for the voltage to become stable. Still here the period of stable signal is long enough for reliable transmission. The speed in this case, as you can calculate, is 115 200 bps. Imagine what would happen if the speed is increased to 1 Mbit per second. The voltage won't rise and fall fast enough to form something resembling a square wave. Nor it will be decodable. Long rising and falling times are caused by capacitance and inductance of used cable. Both parameters increase with length. Hence RS-232 cables (and transmission cables in general) should be short. The situation can be also improved by more powerful line drivers. The more current they can provide the faster is charging of the capacitance of a transmission cable, hence shorter rise and fall times. In practice the role of RS-232 line drivers play voltage converters which are plugged between a microcontroller and the line. Therefore when you choose voltage converter for your project look not only at number of drivers it has, required supply voltage, and capacitance of accompanying capacitors, but also at speeds it can work with. Take care about good quality cable as well.

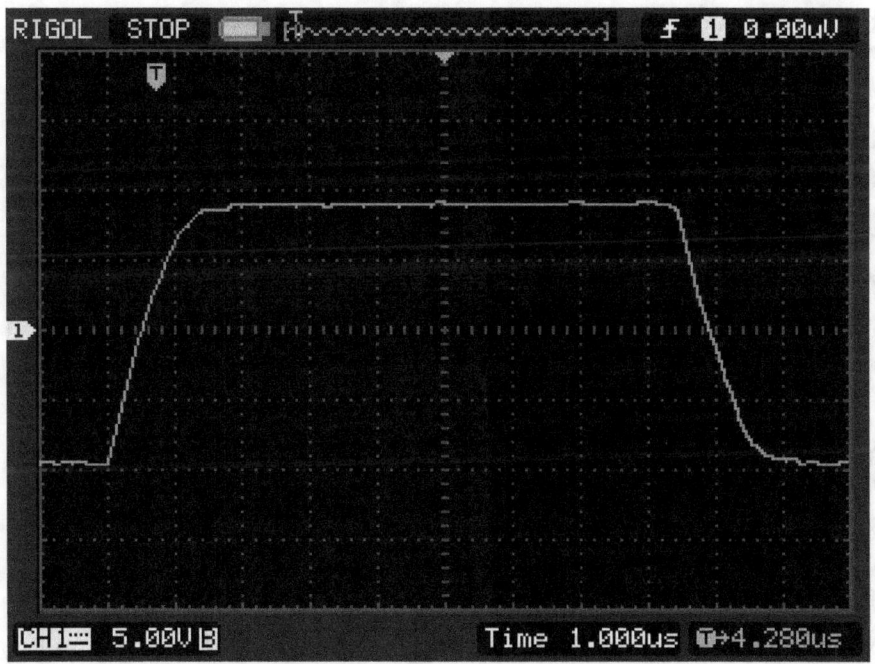

Fig. 1.3.8. Slopes of RS-232 signal at fast transmission. Single bit shown.

1.4. Data encoding

Knowing the basics of RS-232 standard we can ask how data are actually transmitted. As you know, RS-232 is serial data communication standard. This means that data are transmitted bit by bit. In idle state a data line is in low state (logical one). When a device wants to send a byte it first of all sends so called start bit. This lets a receiver know that a byte is about to be transmitted. The start bit is sent by setting logical zero (high voltage level) for a short time on TxD line of a transmitter.

It is then followed by consecutive bits of the transmitted byte, starting with least significant bit first. That's why the bit order may seem to be reversed in respect to how we are used to write binary numbers. Typically there are 8 bits transmitted but it doesn't have to be always true. Rarely you can see other number of bytes transferred: 5, 6 or 7. It must be noted that byte is not always a synonym of eight bits. There were computer systems created which had number of bits lower or greater than eight. Thus sometimes a notion of octet is used for clarity instead of a byte. Data bits may be followed by a parity bit.

If there is a parity bit then it may be set in four different ways: it can indicate even number of ones, odd number of ones, be always one (Mark), or always zero (Space). In the first case the parity bit is set in order to achieve even number of ones transmitted. This includes data bits and the parity bit itself. Therefore if there was odd number of ones in data bits then the parity bit is set to 1 to make the number of ones even. If there was even number of ones among data bits then parity bit is not set because there is already even number of ones. For example for transmitted byte 01010101 the parity bit will be set to 0. In case of odd parity the behavior of parity bit would be reversed. It would be set to one for odd number of transmitted bits, including parity bit itself.

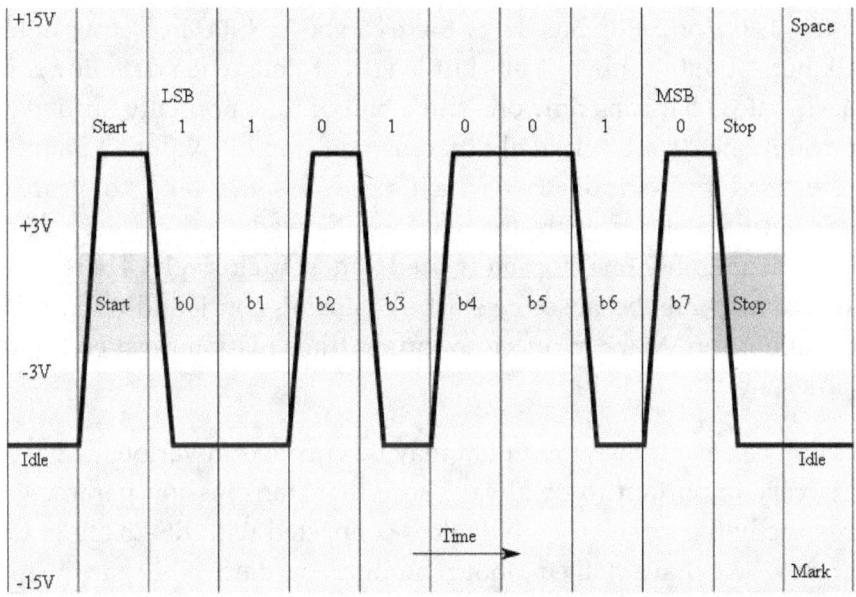

Fig. 1.4.1. Transmission of byte of value 75 (01001011b, 4Bh, ASCII letter 'K') using 8 bits of data, 1 stop bit and no parity bits.

In other words we can say that for even parity the parity bit is set to 1 for odd number of ones in data bits and for odd parity it is set to 1 for even number of data bits. Whether we talk about even or odd parity it is about even/odd number of ones in a byte, not about the byte being even or odd.

Parity bit is a very simple method of detecting transmission errors but it can detect only half of single-bit errors. For Mark and Space parity the error detection is even worse because it allows detecting only errors which occurred during transmission of parity bit.

In many RS-232 applications parity bit is not used. Sometimes this additional bit is used not for error detection but to send additional data. This way 9 bits can be sent instead of 8. The last stage of byte transmission are stop bits which are always logical one (Mark). There can be 1, 1.5, or 2 stop bits. How there can be 1.5 bits? In fact we don't talk here about number of bits but length of time when data line is in Mark state. 1.5 means time one and a half of time normally needed to transmit one data bit. Usually there is one stop bit. Values 1.5 and 2 were used for old devices which needed some time to process received data and prepare for new one. Nowadays there is no such problem and only one stop bit is used almost exclusively. There can't be zero stop bits because the start of transmission is indicated by a transition from Mark to Space so former transmission must end with Mark state.

As you can see transmitted data may be encoded in various ways so it is very important to be always sure that transmission parameters are exactly the same on both devices connected with RS-232 link. Up till now we have talked about number of data bits, parity and number of stop bits. We missed very important thing: speed of transmission. How actually a receiver reads the data sent to it? It must check state of data line in appropriate moments. To do this it has to know how fast data are being sent. Many years ago data rates were very low: 50, 75, 110, 134, or 150 bit per second. Time passed and computers became faster and faster. Higher data rates were introduced: 300, 600, 1200, 1800, 2400, 4800, 7200, and 9600 bps. The last one is now default in many devices. If you don't know RS-232 settings of some device you can start guessing starting with 9600 bps, 8 bits of data, no parity check and one stop bit. The other popular speed is 57 600 bps. This value comes from the fact, that many devices built with '51 microcontroller family use 11.0592 MHz oscillator and due to their architecture they run with 12 times lower speed of 921 600 Hz. In their UART (Universal Asynchronous Receiver and Transmitter) circuits this frequency is divided by 16 which gives 57 600 Hz. Higher data rates which you can encounter are: 14 400, 19 200, 38 400, 57 600, 115 200, 128 000, 230 400, 460 800, and 921 600 bits per second. 115 200 is the highest on many PCs which use 1.8432 MHz oscillators for their UARTs. This frequency divided by 16 gives 115 200 Hz. You may wonder if a non-standard bitrate can be used. The answer is yes, as long as both devices can be configured with the same bitrate value.

State of a line should be read somewhere in the middle of a time of transmission of a bit. We don't want to check state of a line during transition from one bit to the another, we want to read the state when the voltage has already stabilized. Many circuits do this using clock which is 16 times faster than data rate, thus the value 16 mentioned above. For example when speed is set to 9600 bps then UART clock is set to 153600 Hz. When start bit is detected (state changes from Mark to Space, voltage rises from low state to high state) then the receiver waits for 24 UART clock cycles. After first 16 cycles the start bit is finished and during the following 8 cycles receiver waits for the middle of the first bit. When it comes UART reads state of its RxD line and remembers it. Then it can wait another 16 cycles for the middle of the second bit. Some UARTs check state of the transmission line three times during transmission of a bit and then compare read values. When one is different from other two, the value from those two reads is taken as the right one. This is the way of error detection and correction which may be needed in high noise environments.

Thinking about data rates and about reading them in time slots we may start to wonder what happens is transmitter and receiver use slightly different clocks. If there are separate devices using own clocks then after some time they would desynchronize and receiver would start to read data in wrong moments. Fortunately it won't happen because both devices are synchronized with every start bit. So even if there is a slight difference it won't affect correctness of a transmission. But keep it under about 3%.

1.5. Flow control

As you know, there are CTS and RTS lines. When idle the RTS line is in logical 1 state (low voltage, about -12V). If DTE wants to send a data, it sets RTS to logical 0 (high voltage, about +12V). Then DCE knows that a transmission is about to start and when it is ready it also changes state of CTS line from 1 to 0. At this moment DTE knows that it can start actual data transmission. Both RTS and CTS use the same voltages as RxD and TxD lines. Thus if you want to connect them to a microcontroller you need voltage converter just like for RxD and TxD. Fortunately MAX3232 has two pairs of drivers/inverters so there is no problem with connecting all four lines. This is hardware flow control.

There can be also software flow control; it's called XON/XOFF. XON means transmit on and XOFF means transmit off. It is realized by sending special bytes. XON is encoded by a byte of value 17 (0x11 in hexadecimal system) which is an ASCII character named DC1. On the other hand XOFF is encoded by a byte of value 19 (0x13 in hexadecimal system) which is an ASCII character named DC3. When a receiver got too many data it can send XOFF byte to a transmitter so it can suspend transmission. When the receiver is ready to receive further data it can send XON byte and receive data again.

As you can see we have here two different approaches. Hardware flow control is easy but requires two additional lines. In case of software flow control now additional lines are needed. Unfortunately we have to transmit additional bytes which can be inconvenient. It can slow a transmission down and causes problems when binary, arbitrary data are sent. If a byte 17 or 19 is transmitted it's not known whether it's data bye or flow control byte. Used transmission protocol must take such situation into account and assure that received bytes would be unambiguously decoded. In both cases you can observe that in flow control there is only one party which controls the flow.

In many cases no flow control is used. It also applies to examples presented in this book.

1.6.   *Serial port USB adapters*

Many computers don't have RS-232 ports now. Consumer devices use mainly USB standard and with time RS-232 became obsolete. But if we want to build devices controlled with RS-232 we need such port in our computer. If we don't have it, we need some adapter. The most popular ones are cheap USB <-> RS-232 adapters. They have form of a cable with DE-9F plug on one end and USB plug on the other end. The DB-9F plug is big enough that it can contain also the conversion between RS-232 and USB. When such cable is plugged to a computer, it is seen as virtual serial (COM) port. Of course appropriate drivers are needed. They are usually provided with the adapter. If they are not, we have to find them on the Internet. If we can't find driver using name of our adapter or we don't know neither model nor manufacturer we can search using VID and PID (vendor ID and product ID) of the device. When it is plugged in then it's displayed in device manager in Windows. We can open properties window and find both identifiers and use them to find drivers on the Internet.

Fig. 1.6.1. Popular RS-233-to-USB adapter cable based on PL-2303 chip.

Sometimes it may happen that we have downloaded the drivers but Windows can't find them although a proper folder with driver files is selected. In this case first of all you have to look for another drivers. If you can't find any you have to use the downloaded ones. How to do it? Windows looks for VID and PID values in INF file. If they don't match, Windows doesn't install the driver. So it is required to replace VID and PID values in INF file with values obtained from device manager. When it is done, Windows should use drivers placed in selected directory. However, if the drivers didn't match, you must be aware that they are probably designed for slightly different version of the device and they can work improperly.

1.7. Connectors, cables, and connections

A cable which is intended to be used for RS-232 transmission needs to have all necessary wires. If hardware flow control is not needed, just three wires are enough: two for sending and receiving data and one as common ground. In case of utilizing hardware flow control more wires are required, up to nine. You have to choose a cable suitable for your needs. If you buy one, make sure it has all the lines you want.

Of course bunch of wires can do the job but adding some connectors on both ends would make life much easier. Typical ones are DB-9. Their correct name is actually DE-9 but the DB-9 name is so widespread that we stick to DB-9 in this book. These connectors have 9 pins which is exactly what is needed to transmit all RS-232 signals. As many other connectors, DB-9 is available in two variants called genders. DB-9F is a female connector, the one with holes. On the other hand a DB-9M connector has pins and is called male. In general DTE devices, such as computers, are equipped with male connectors and DCEs have female connectors. However you can encounter some exceptions, therefore it's good to check pinouts before connecting unknown devices. In any case for a device with a socket of given gender we obviously connect plug of opposite gender.

Fig. 1.7.1. DB-9 plugs: male (left) and female (right).

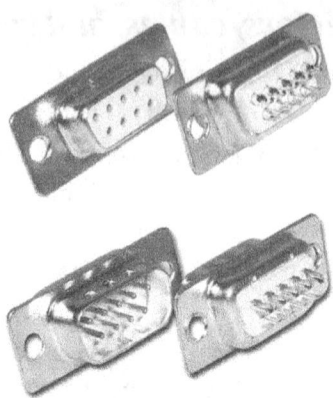

Fig. 1.7.2. Female and male DB-9 connectors, connecting side and cable soldering side.

If you already have a cable it may turn out that you can't use it because genders are not matching. For example both devices have male connectors but your cable has male connector on one end and female on the other. In such case you can use so called gender changer. It's very simple device consisting of two connectors of the same gender. But be careful. In any case first of all you have to determine what kind of cable you have. Usually it's a regular RS 232 cable (male – female) or a null-modem cable (female – female) used for DTE – DTE connections. Null-modem cable has different pin connections and can't be changed into regular one with just a gender changer on one side.

Fig. 1.7.3. Gender changers: female-to-male and male-to-female

DB-9s are not the only connectors used for serial cables. In old equipment you can find wider connectors having 25 pins. They are called DB-25. In modern devices some manufacturers like to use RJ-45 connectors as they are smaller and easier to connect and disconnect. Unfortunately there are many pinouts for RJ-45 so you have to check what the pinout of the device you are dealing with is. The pinout defines by EIA/TIA-561 is shown on fig 1.7.5.

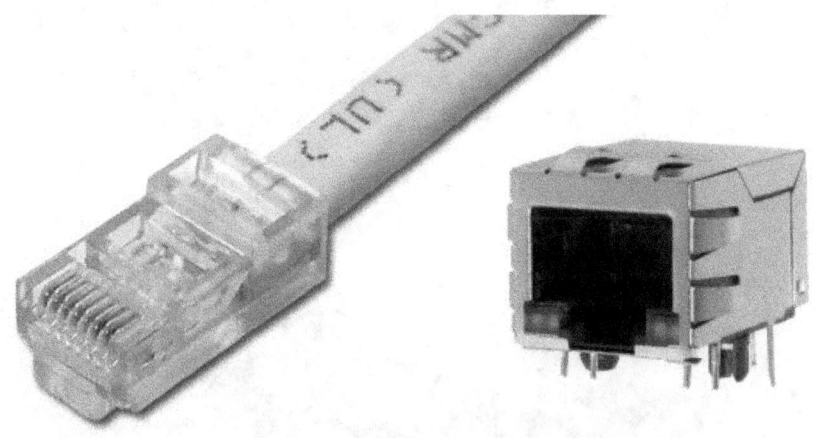

Fig. 1.7.4. RJ-45 plug with cable and PCB-mounted socket

RJ45 pin no.	Name
1	DSR/RI
2	DCD
3	DTR
4	GND
5	RXD
6	TXD
7	CTS
8	RTS

Fig. 1.7.5. RJ-45 pinout according to EIA/TIA-561 standard.

Serial port and microcontrollers

We also have to mention DB-25 connectors. They were popular in the past but currently it's hard to find such connectors in RS-232 devices. The additional 16 pins, in comparison to DB-9, were usually left unconnected; sometimes they were used for other signals. To connect two devices, from which one is equipped with DB-9 and the other with DB-25 connector, we can use a cable with appropriate plugs or an adapter.

Fig. 1.7.6. DB-25-to-DB-9 adapter.

OK, we know now what kind of connectors we need. We have also learned what the RS-232 lines are. But how to connect them? Let's look at table. 1.7.7. It shows connection between DTE and DCE. DTE has male socket and DCE has female one. We connect them with simple female-male cable which just connects pins of the same numbers. As you see signals are described from the point of view of DTE.

DTE device (computer)   DCE device (modem)

Pin	Signal name		Signal Direction	Pin	Signal name	
1	Carrier Detect (DCD)	CD	⟵	1	Carrier Detect (DCD)	CD
2	Receive Data (RxD)	RD	⟵	2	Receive Data (RxD)	RD
3	Transmit Data (TxD)	TD	⟶	3	Transmit Data (TxD)	TD
4	Data Terminal Ready	DTR	⟶	4	Data Terminal Ready	DTR
5	Signal Ground	GND	—	5	Signal Ground	GND
6	Data Set Ready	DSR	⟵	6	Data Set Ready	DSR
7	Request To Send	RTS	⟶	7	Request To Send	RTS
8	Clear To Send	CTS	⟵	8	Clear To Send	CTS
9	Ring Indicator	RI	⟵	9	Ring Indicator	RI

Fig. 1.7.7. DTE and DCE devices connected using DB-9 connectors

And here is a little confusion. For example pin 2 of DTE is connected with pin 2 of DCE. Both pins are called RxD but only DTE's pin receives data, DCE's pin transmits data. This is somewhat counterintuitive, therefore sometimes DCE's pin 2 is called TxD and pin 3 is called RxD. This goes further and also CTS is replaced with RTS and DSR with DTR. While with TxD and RxD it can be understood, in case of the remaining four control signal it's not correct. CTS is not an input for RTS, it's a separate signal sent as an acknowledgement for RTS. The same applies to DSR and DTR. Following paragraphs describe DTE-DTE and DCE-DCE connections which should clarify the problem.

Above we are talking about DTE and DCE, however on the beginning of the book it was said that the distinction for DCE and DTE devices goes away. That's right, not so rarely we want to connect two computers or two microcontroller devices and we can't say which one is DTE and which one is DCE. In such a case we use so called null-modem which in fact is just a cable with connections made in a specific way. If we don't use handshaking we just need three wires: TxD, RxD and GND. Note, that we use two devices of the same type: DCEs or DTEs. Therefore we can't connect their pins in a standard way. If we have two DTEs then on both sides pins 3 are outputs and pins 2 are inputs. So we have to connect pin 3 (TxD) of one DTE with pin 2 (RxD) of the other DTE and pin 2 of the first DTE to pin 3 of the second DTE. Of course pins 5 (GND) have to be connected together. The same applies to two DCEs.

If handshaking is used we have two cases: we use real full handshaking or make a loopback. In the first case we need wires for handshaking lines. CTS of one device has to be connected with RTS of the other and vice versa. DSR and DTR are connected in analogous way. The RI signal is left unconnected: there is no telephone line so there is no ring signal.

Pin	Signal name		Signal Direction	Pin	Signal name	
1	Carrier Detect (DCD)	CD		1	Carrier Detect (DCD)	CD
2	Receive Data (RxD)	RD		2	Receive Data (RxD)	RD
3	Transmit Data (TxD)	TD		3	Transmit Data (TxD)	TD
4	Data Terminal Ready	DTR		4	Data Terminal Ready	DTR
5	Signal Ground	GND		5	Signal Ground	GND
6	Data Set Ready	DSR		6	Data Set Ready	DSR
7	Request To Send	RTS		7	Request To Send	RTS
8	Clear To Send	CTS		8	Clear To Send	CTS
9	Ring Indicator	RI		9	Ring Indicator	RI

Fig. 1.7.8. Null-modem connection for two DTEs (triangle arrows) or DCEs (line arrows).

Now you may ask why we connect RTS with CTS and DTR with DSR if, as it was said earlier, CTS is not an input for RTS and DSR is not an input for DTR. That's because a null-modem is a special case and there is actually no choice. If there are two DTEs then their RTS pins are outputs and their CTS pins are inputs. By connecting RTS pin of one DTE with CTS of the other DTE and CTS of first DTE with RTS of the second we can make RTS/CTS handshaking work although not with 100% accordance with the RS-232 standard. A DTE here gets CTS signal not as a response for own RTS signal but when the other device sets its RTS signal which may happen after or before the first DTE sets its RTS signal. Similar situation is with DTR and DSR lines. However you may find circuits where DSR is connected also with DCD of the same device. This way the second can simulate its readiness and carrier signal for the first device at the same time. But be careful as this works only for two DTEs. If we deal with DCE devices then both DSR and DCD are outputs and they can't be connected together. Anyway usually there is no need to connect DCD to anything in a null-modem cable as this signal is typical for modems working with telephone line.

Pin	Signal name		Signal Direction	Pin	Signal name	
1	Carrier Detect (DCD)	CD		1	Carrier Detect (DCD)	CD
2	Receive Data (RxD)	RD		2	Receive Data (RxD)	RD
3	Transmit Data (TxD)	TD		3	Transmit Data (TxD)	TD
4	Data Terminal Ready	DTR		4	Data Terminal Ready	DTR
5	Signal Ground	GND		5	Signal Ground	GND
6	Data Set Ready	DSR		6	Data Set Ready	DSR
7	Request To Send	RTS		7	Request To Send	RTS
8	Clear To Send	CTS		8	Clear To Send	CTS
9	Ring Indicator	RI		9	Ring Indicator	RI

Fig. 1.7.9. Null-modem with loopback (triangle arrows for configuration with two DTEs, line arrows for configuration with two DCEs).

We can use loopback when our devices require handshaking but for some reason we don't want to provide wires for it. Then we can cheat: connect RTS and CTS together, and DSR and DTR. This way the device will get CTS signal from its own RTS line, and DSR signal from its DTR line. If we want to use RTS/CTS handshaking but don't need DTR/DSR then we can make loopback just for RTS and CTS.

1.8.   Built-in USB interface

If we want our device equipped with RS-232 interface to work with a computer which lacks this port, we can go the other way around. Instead of using an adapter which is external from the point of view of our device, we can put such converter inside. We just need some little space on our PCB for FT232 or similar integrated circuit. It should be connected to serial port pins of a microcontroller used in our device and to the USB socket. That way we can connect our device to a PC with an ordinary USB cable. As in previous approach, the device will be seen by operating system as a virtual serial port. This solution is very popular, many USB devices are actually RS-232 devices equipped with some IC from FT232 family. This is because USB is very common now but quite complicated at the same time. It is often less expensive and time-consuming to add serial-to-USB converter than to create pure USB device.

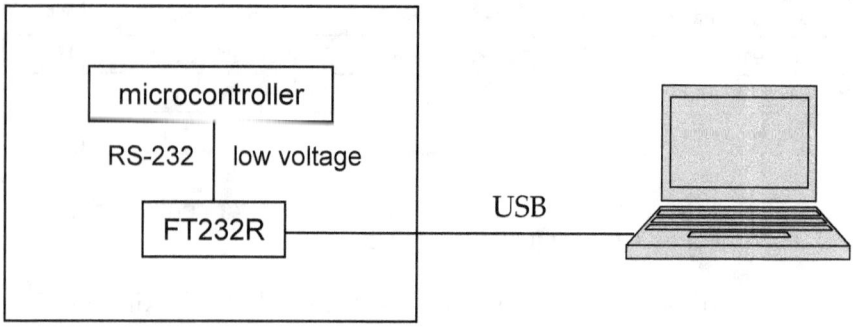

Fig. 1.8.1. Using FT232R to add USB capability to a microcontroller device

FT232 is a popular family of USB-to-UART interfaces which can be connected directly to a microcontroller or voltage converter for RS-232/RS-422/RS-485. Older chips from this family required external components, e.g. EEPROM for storing configuration. One of the newest chips, FT232R, doesn't require them and is very easy to use. It is available in two versions which differ in package: FT232RQ has 32-

pin QFN package and FT232RL has 28-pin SSOP package. It can be powered from the USB bus or from another power source. It has internal 3.3V voltage regulator which you can use if the rest of your circuit can work with such supply voltage and doesn't consume more than 50 mA. FT232R offers hardware flow control lines and has five I/O lines which can provide alternate functions according to the configuration stored in internal EEPROM. User can also change chip's VID and PID which is useful if we want FT232R to be seen as our custom device.

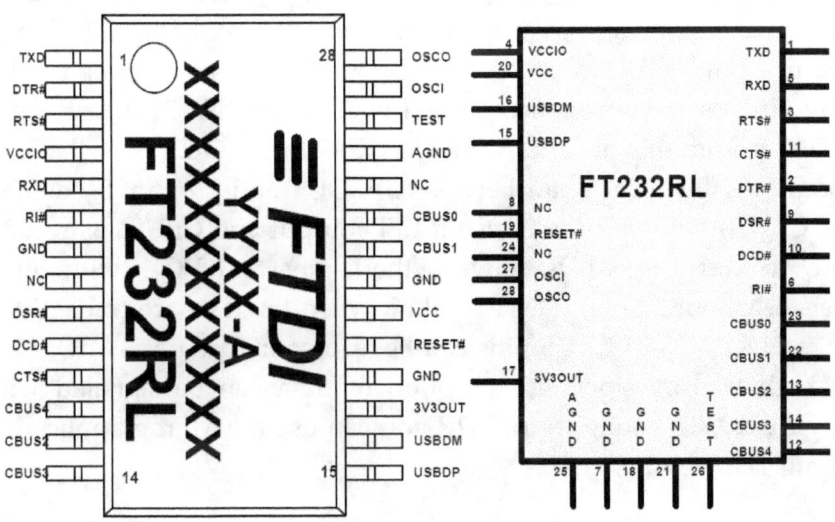

Fig. 1.8.2. FT232RL package and symbol

It's quite easy to use FT232R. USBDM and USBDP are USB interface D- and D+ pins and should be connected to a USB host. In practice this usually means they should be connected to appropriate pins of a USB-A plug (male), USB-B socket (female), or USB mini-B plug (female). It depends whether you are going to plug your device directly to a USB port (like a flash drive) or use a cable. All three GND pins, AGND pin, and TEST pin should be connected with USB ground. VCC, VCCIO and 3V3OUT should be connected depending on how your device is powered and what is the value of its supply voltage. VCC pin provides power for chip's core. It can be from 3.3V to 5.25V. This means that you can use the voltage provided by USB interface and connect VCC line to USB VBUS line. If for some reason you don't want to draw power from USB interface, you have to provide supply voltage from other source. VCCIO is a pin which provides power for FT232R peripheral pins, including serial and I/O (CBUS) pins. Its voltage should have the same value as supply voltage of a circuit connected to those pins. Therefore for USB-powered designs and 5V circuits you can connect both VCCIO and circuit's power line to USB VBUS. 3V3OUT pin is a nice feature for 3.3V circuits. If your device doesn't need more than 50 mA you can power it from this pin and connect VCCIO to 3V3OUT as well. There should be 100nF capacitor between 3V3OUT and ground. By default the CBUS0 pin works as transmit LED output and CBUS1 as receive LED. If you plug LEDs between these pins and VCC (with serial resistor, about 270 Ω) they will blink when the data are transmitted or received. On of CBUS pins can be also configured as TX/RX LED so both transmission and reception of data can be signaled with single LED. To configure FT232R you can use the MProg application from FTDI website.

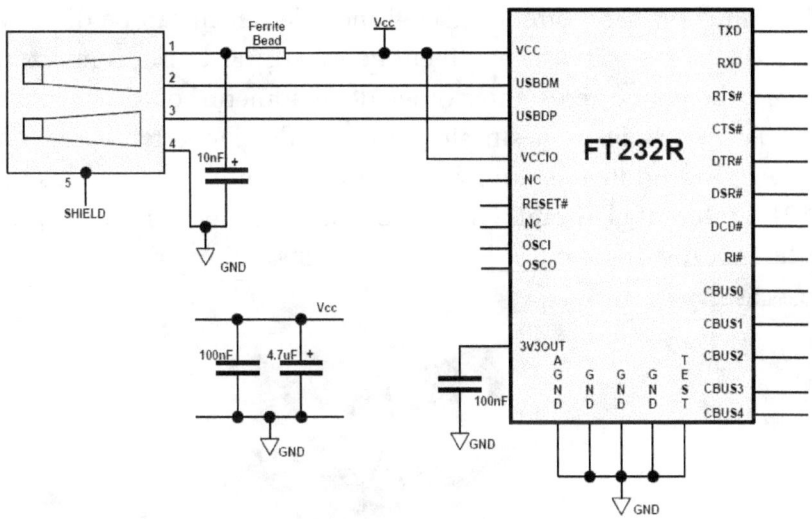

Fig. 1.8.3. USB bus powered confguration

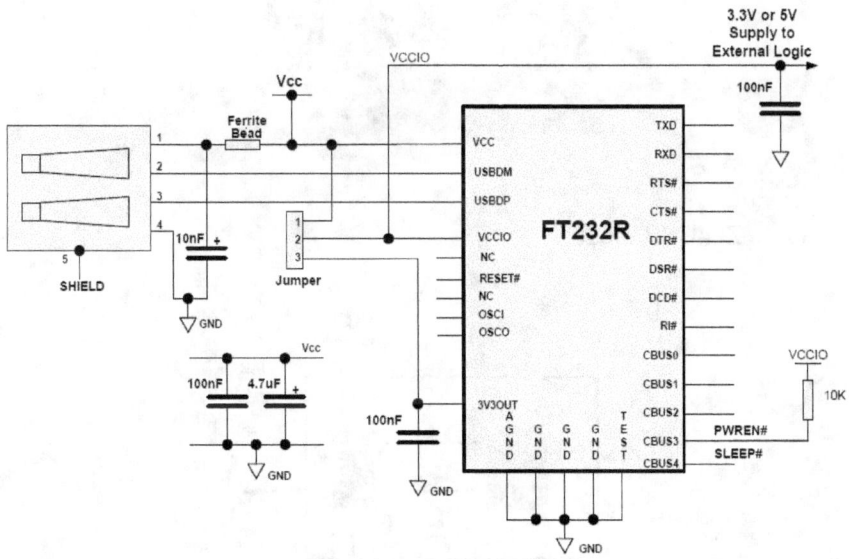

Fig. 1.8.4. Configuration where the rest to of the device can be powered from USB bus (5 V) or FT232R's internal voltage regulator (3.3 V).

41

Serial port and microcontrollers

The Kamami company offers ZL4USB module which can be directly connected to some of its development boards using 14-pin connector. It can also be connected to any other development board or circuit with cables. ZL4USB is a simple serial<->USB converter based on FT232R. Its additional feature is galvanic isolation realized using ISO7221A integrated circuit. Only TxD and RxD lines are accessible with the module. It's kind of limitation but should not be a problem in most cases.

Fig. 1.8.5. ZL4USB module

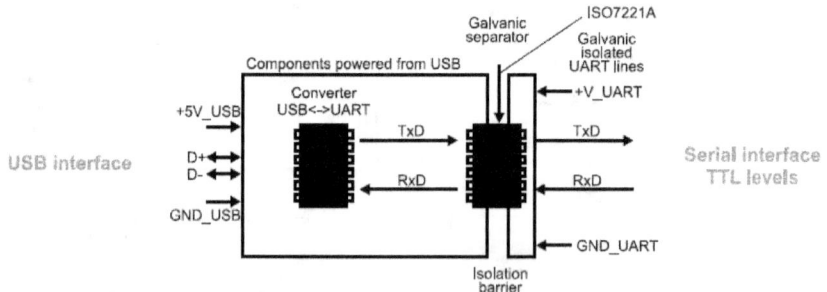

Fig. 1.8.6. ZL4USB block diagram

Although FT232 is a very popular way of connecting RS-232 devices to USB ports, there is a project which can sometimes work as a good and cheaper replacement. It is called CDC-232 and is described at http://www.recursion.jp/avrcdc/cdc-232.html The idea is to use cheap AVR microcontroller to work as serial-to-USB converter. In the simplest version you need just ATtiny45, LED, five resistors, a capacitor and USB connector. Such circuit can work with speeds up to 4800 bps. A more advanced version is based on ATtiny2313 and can work at up to 38400 bps. If you use ATmega8/48/88 you can have also the RTS/CTS flow control. CDC-232 is a very simple and inexpensive way to have virtual COM port and add USB capability to serial device. It works on Windows, Linux and Mac OS X. It has of course also some limitations, for example it can have problem with working with virtual machines, like VMware. Pay also attention to resistors R4 and R5 connected to TxD and RxD lines. They are necessary if a device working with CDC-232 is powered from different power source, which is a frequent situation. The CDC-232 uses V-USB which is software USB stack which allows using USB on AVR chips which don't have built-in hardware USB controller. More about V-UBS you can find at http://www.obdev.at/products/vusb/index.html

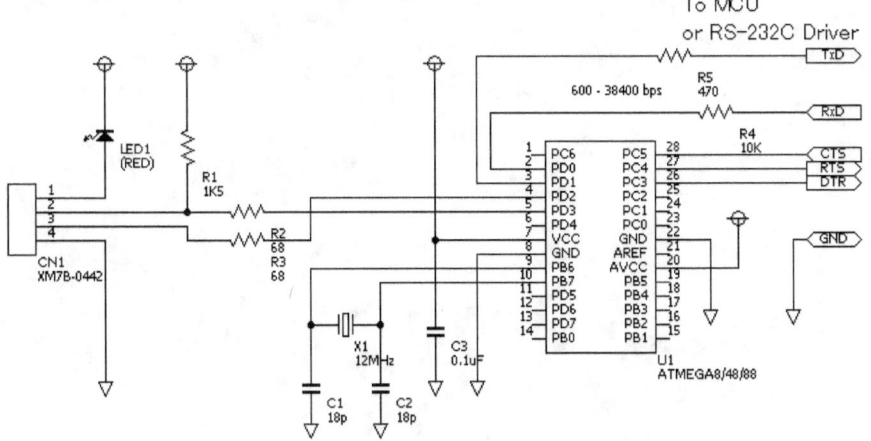

Fig. 1.8.7. CDC-232 schematic for ATmega8/48/88-20

Fig. 1.8.8. CDC-232 with ATmega88PA assembled as an external device (prototype, without resistors on RxD and TxD lines)

2. The RS-485 standard

As you know from previous chapters, the RS-232 standard uses one wire (plus common ground) to transmit data in one direction. This is a very simple approach but has also some disadvantages. One of the problems with RS-232 is low speed and another one is short link length. They are caused by the fact that single data line is prone to receiving electromagnetic noise which induces additional voltage on the line. If the noise has enough power and both a transmitter internal resistance and a line resistance are high enough then the voltage on the input of a receiver is going to have such a value that it will be interpreted as a logical state different than the one sent by a transmitter. The higher the noise the shorter links and lower data rates have to be used. Speed is also limited by capacitance of wires.

The other problem with single data line is the common ground line. If it is long, it can have different voltages on the ends, especially if high currents are conducted. Voltage difference on ground line can cause the receiver to misread data.

RS-232 can be tricky because of used voltages too. The driver in a transmitter has to supply a voltage at least as low as -5 V and as high as +5 V. This is problematic because majority of devices are powered from single voltage, often lower than 5 V. Although there are many inexpensive voltage converters the issue remains.

2.1. Differential line

RS-485, known also as IA-485, tries to solve above problems. Instead of single-ended, unbalanced line it uses differential, balanced line. There is a pair of wires for each data link plus common ground. Data are not transmitted by means of voltage difference between single line and ground but as a difference of voltages in a pair of wires conducting current in opposite directions. When a noise induces voltage, it's usually similar value in both wires of a pair. Thus the difference remains the same and a receiver can correctly read transmitted data. The noise is cancelled or much reduced. It must be noted however, that voltages in differential link are still measured with respect to ground. The signal is the difference between the voltages anyway.

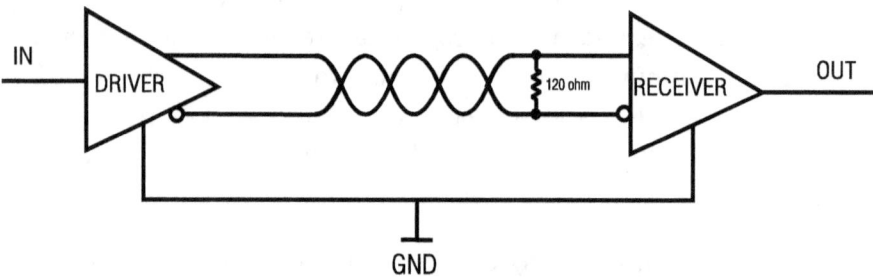

Fig. 2.1.1. RS-485 differential line with driver, receiver and terminating resistor

To talk about voltage difference we need to know what is compared with what. One line in a pair is called A and the other one is called B. Logical one is transmitted when B is more positive than A, logical zero is transmitted when B is more positive than A. However when interfacing digital systems we again, like in case of RS-232, deal with inversion. Consider an RS-485 receiver, e.g. MAX485. When logic one is present at chip's input, it makes output A more positive than output B. Hence there is logic zero on the RS-485 link. Analogous situation applies to logic zero at the input. Therefore we have an inversion of logic states. On the other end of an RS-485 link there is second MAX485 chip. When it has higher voltage on A input than on B input then it produces logical one at its input. As we can see there is double inversion and logic levels on input of first chip and output of second chip are the same.

As you can see in typical applications, where specialized ICs are used, there is no problem with inverted logic signals. In practice another problem is more serious: different naming conventions. Some manufacturers define A and B in opposite way. Some use "+" and "-" labels and some introduce "Signal" and "Inverted Signal" convention. Usually "+" is the same as B which is the same as "Signal", likewise "-" is the same as A which is the same as "Inverted Signal". Because of this inconsistency you have to check what rule was chosen by the manufacturer of you device or integrated circuit. However you don't need to worry much. If your connection is not working try the opposite. In case of incorrect connection nothing is going to be broken, the data are just going to be inverted.

2.2. Data encoding

RS-485 doesn't define any data transmission protocol, it just focuses on electrical characteristic of a link. The way of sending ones and zeros is defined but it's up to a user to define such things as data rate, dividing data into chunks (e.g. bytes), marking start and end of transmission or error detection. Usually RS-232 protocol is used but you may come up with own ideas. In case of more than one transmitter on a link you have to take care about collision detection. Anyway all connected devices have to follow the same rules of communication.

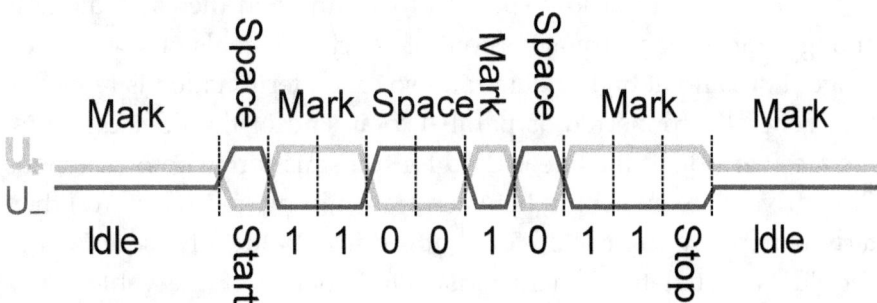

Fig. 2.2.1. A byte transmitted over RS-485 link using RS-232 encoding. Image source: Wikipedia.

2.3. Electrical characteristic

Contrary to RS-232, the RS-485 doesn't need high voltages to transmit data. On a receiver's side required voltage difference may be as low as ±200 mV. If voltages differ by less than 200 mV then the state is undefined. A transmitter has to provide minimum ±1.5 V difference. Maximum difference is ±6 V. On both sides voltages of each link in a pair can't exceed –7..+12 V voltage range in respect to common ground.

According to the standard there can be up to 32-nodes connected to one RS-485 link. Input resistance of each one of them has to be not less than 12 kΩ. The more nodes the lower resulting resistance. Transmitter has to have output resistance low enough to guarantee required voltage difference during data transmission (±1.5 V). By increasing input resistance you can increase number of nodes or lower power consumption. Higher resistance means lower bit rate though. Speed of a link is also determined by cable length. RS-485 allows achieving bit rate up to 35 Mbps at 12 meters and 100 kbps at 1200 meters which is greatest length of a link in the standard. Typical cabling is a twisted pair of wires, having characteristic impedance equal to 120 Ω.

2.4. Terminations

On short links and at low speeds of transmission the cable alone is enough for proper transmission of electric signals. However on longer lines and at higher data rates so called termination is required. It is generally connected as parallel to an end of the RS-485 link, at the receiver side. If the line is used for transmission in both directions (line drivers are switching between transmitting and receiving) then termination are connected on both sides. Value of its resistance should be equal to a characteristic impedance of the cable. Thus usually we have one or two resistors of value 120 Ω. Anyway you should check the characteristic impedance when you buy cable for your RS-485 line or measure it.

We need terminations because of the fact that transmitted signal is reflected from ends for the transmission cable. This unwanted effect can cause errors in data because reflected signal interferes with the proper one. Knowing that reflected signal damps out with time we had to make sure it's value will be low enough after the time needed to transmit half of a data bit. You can easily observe now that the problem becomes more serious for higher bit rates and longer links. The higher the speed the shorter is time intended for transmission of one bit. The longer a link the more time a reflection needs to damp out.

To determine if terminations are necessary you have to make a few simple calculations. First of all let's check how long does it take for a signal to travel through a link and go back. Speed of electric current is sometimes compared to speed of light. But as a light is slower in a medium than in vacuum the electric current is slowed down in metal. We may assume that speed of current is 2/3 of speed of light which gives us $2*10^8$ m/s. Let's assume that length of our link is 100 meters. When traveling forth and back it gives 200 meters. Thus the electric current needs 1 microsecond for 1 cycle. Let's assume we need four cycles for a reflection to dump out to a safe enough level. Therefore we need 4 microseconds. Typically a value of transmitted bit will be read in the middle of the time it takes to transmit this bit. So this time need to be at least 8 microseconds. This gives us final, maximal data rate which is possible without terminations: 125 kilobits per second. The standard, close value is 115 200 bps. For higher data rates at given link length (100 m) we should use terminations, for lower ones we don't have to.

2.5. Full duplex and half duplex

The RS-485 data link is unidirectional at any given time. We can't simultaneously transmit and receive data having only one link (pair of wires). If we have several nodes on a link and at least some of them need to both send and receive data we can use two approaches: half duplex and full duplex.

In half duplex mode there is just one data link and all the nodes are connected to it. Usually we use specialized chips for interfacing digital circuits with RS-485 line. Such chips often work as transceivers, having built in transmitters and receivers. Transmitter can be put into three states: zero, one, and high impedance. In one and zero states transmitter works as a driver which forces a voltage difference between two wires in the data link. In high impedance state the driver doesn't try to set any voltages, it is seen as disconnected. This allows other node to send data over the link. In other words during transmission only one node is active, the rest have their transmitters disabled. An RS-485 transceiver generally has a pin which can enable or disable built-in transmitter. It is usually called DE which means Driver Enable. This pin can be controlled by a microcontroller which can use it to enable transmitter when it is about to send data and disable it during transmission performed by other nodes. There is also equivalent pin called /RE which is negated Read Enable. It is used to turn on chip's built-in receiver. In many applications both control pins are connected together. Thus turning on the receiver turns off the transmitter and the other way. As we can see a microcontroller can use one pin to control direction of transmission. Half duplex is often used when there are many nodes attached to a link. This way only one link is needed to connect all the nodes. The disadvantage is that there can be only one transmission at a time and there must be special algorithm implemented to avoid collisions when two or more nodes want to transmit data.

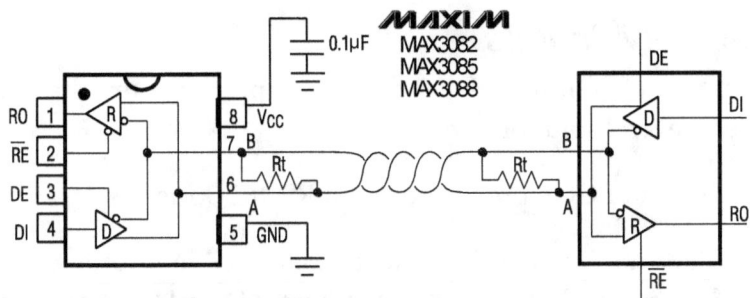

Fig. 2.5.1. Maxim RS-485 transceivers with /RE and DE pins for enabling driver and receiver blocks; RO – receiver output, DI – driver input

On the other hand, in full duplex mode there is no switch of direction of transmission. There may be many nodes attached to a line, both transmitters and receivers, but they role doesn't change in time. This somewhat simplifies transmission because nodes don't need to care about switching direction of transmission. On the other hand it requires more links which means more wires. And if there is more than one link still there is problem of avoiding collisions.

2.6. Adapters

PCs don't have built-in RS-485 ports. If support for this standard is required, several solutions are available:

- PCI and PCI Express cards with RS-485 ports

  They are mounted inside a computer and provide many features such as ADDC (Automatic Data Direction Control) for half duplex lines, optical insulation, terminations, and overvoltage protection

- USB <-> RS-485 adapters

  Usually they are based on FT232 chip family which are USB<->RS-232 (TTL) converters so the adapters is seen as a COM port in a computer. These converters are connected to RS-485 drivers which makes complete USB<->RS-485 adapter

- RS-232 <-> RS-485 adapters

  They allow you to turn an RS-232 port into RS-485 port. Usually they offer one, half duplex line and are powered from serial port. In many cases they consist of RS-232 to TTL voltage converter and RS-485 transceiver.

In any case you have to refer to the data sheet of particular device to check what functionality it does offer.

Fig. 2.6.1. PCI card with RS-485 interface

If you want to build own RS-232 to RS-485 adapter you can you the same approach as in devices available on the market: RS-232 to TTL voltage converter attached to TTL to RS-485 converter. As a chip working on RS-232 side you may choose MAX3232 or similar IC. For the RS-485 side you have to think about mode of operation: half duplex or full duplex. For full duplex you need a chip with separate driver and receiver which can be connected to two different RS-485 lines. This way sending and receiving data can be done at the same time. A popular example of such an IC is MAX490. In case of half duplex required, another chip is needed. It needs to have a driver and receiver as well but working on the same RS-485 line. Another requirement is a possibility of choosing data transmission direction. This is fulfilled in MAX485 integrated circuit.

If you want a really cheap and simple RS-232 <-> RS-485 full duplex connection you can make it with just two resistors. Just connect B wire from RS-485 link (designed for incoming data from RS-485 device) to RxD wire of RS-232 link. Leave A wire unconnected. For second RS-485 line (which we use to send data to a RS-485 device) connect A wire to common ground. B wire should be connected to TxD pin of RS-232 port through a voltage divider because RS-232 voltages exceed maximum values allowed for RS-485. This way, using just two resistors, we can connect an RS-232 line with full duplex RS-485 line. You must remember however, that this is a non-standard solution and can work only in some circumstances. One problem is that voltages on RS-485 link may be not correctly interpreted by an RS-232 receiver. An RS-485 driver needs to provide appropriate voltage difference between A and B wires but RS-232 receiver would check a difference between B and GND. We can imagine a situation when voltages on A and B lines during data transmission are always negative with respect to common ground and it will be correct from the point of view RS-485 standard. An RS-232 receiver will interpret these voltages always as logical low though. On the other hand RS-485 drivers are often powered from +5V supply and they work by providing +5V or 0V voltages on A and B lines. +5V is a correct RS-232 voltage level but 0V lays in a forbidden zone. Fortunately RS-232 receivers usually interpret voltages below +1V as logical low state so such a setup can work. The other problem with described minimalistic circuit is lack of terminations. Therefore it can work only on short distances and lower speeds.

3. AVR microcontrollers

One of the two groups of microcontrollers used in sample circuits presented in this book are the ones belonging to the AVR family. They are 8-bit RISC microcontrollers developed by Atmel and are based on modified Harvard architecture. The AVR ICs are very popular among hobbyists because of low cost and easy programming. They have built-in flash memory for program storage and offer many useful peripherals. AVRs can be programmed while they are mounted in target device. It is called In-System Programming (ISP). There is no need to remove them from socket and insert into a programming device, like it is necessary for same microcontroller families. Thus AVRs are easy to program and debug. Programming is usually performed through SPI interface (Serial Peripheral Interface Bus) or JTAG interface.

AVR microcontrollers require almost no external elements. They have internal RC oscillator which provides clocking signal. If greater clock accuracy is needed, one can connect additional crystal oscillator with capacitors. The other common external add-on is basic RC circuit providing reset pulse on power-on.

There are many software packages for programming AVRs, including commercial BASCOM AVR and free WinAVR. AVR microcontrollers are divided into three main groups: tinyAVR, megaAVR and XMEGA. The first one offers 1 – 8 kB of flash memory and has only basic peripherals. It also has low power consumption and is very inexpensive. megaAVR has more peripherals, bigger flash memory (4 – 256 kB) and richer instruction set. In case of XMEGA microcontrollers we get even bigger memory and additional features like DMA (Direct Memory Access) and hardware support for cryptography. There are also other AVR ICs, designed for special purposes.

Fig. 3.1. ATmega32 in 40-pin DIP package and ATmega8L in 32-pin TQFP package (not in scale).

3.1. Software tools

In order to create software for microcontrollers we need some tools: editor, compiler, debugger, and programmer. Let's take a look at some useful applications.

3.1.1. WinAVR

WinAVR is a suite of open source development tools for the AVR microprocessors hosted on the Windows platform. It is based on GNU GCC compiler for C and C++. The biggest advantage of WinAVR is that you can create software, including commercial programs, with no cost. If you are more advanced programmer you can also modify the included tools. The downside of WinAVR is more complicated usage. It doesn't come with any nice integrated development environment. There is just Programmers Notepad provided, which is not so bad after all. You can use this one or any other editor which you consider suitable for your needs, for example PSPad, Notepad++, UltraEdit-32 or Vim for Windows.

Compilation can be done step by step, issuing commands for each compilation step, or with a Makefile file processed by the make command. Of course the latter is much more convenient. You can launch compilation from command prompt (cmd.exe) or, if there is such a function, from you editor. When the code is compiled you have to send it to a microcontroller using ISP or JTAG interface. WinAVR comes with very good programming tool called AVRDUDE which supports many hardware programmers and microcontrollers.

Most of the tools included in WinAVR are text mode programs, running inside command prompt (cmd.exe or powershell.exe). To be available from any directory, their location must be listed in an environment variable called *PATH*. It is done by WinAVR installer which adds *bin* and *utils\bin* directories to *PATH*. In typical cases you don't have to pay attention to it but in case of problems this information can be useful. There are two common causes of problems: there is no path to WinAVR utilities present in *PATH* variable or you have a few executables of the same name in different places on disk. This second situation happens usually when you have other development environment installed in your system. Then you can have *make.exe* from WinAVR in one place and *make.exe* from the other development environment in other place. Depending on order of paths in *PATH* variable it may turn out that you call incorrect *make.exe* when you type *make* in command prompt. In such a case change order of paths or create a CMD script which will set *PATH* variable correctly.

If you need to have a command prompt with a particular set of environment variables or you want to have some command executed automatically when starting command prompt, you can use following idea. Create a CMD script (batch file) with all commands you want to be executed. Create a shortcut to command prompt (cmd.exe). At this moment you have an ordinary shortcut to command prompt. To have desired commands executed, you have to edit it. Open properties of the shortcut and find a text box containing target path. At the end, add /k parameter and path to your script. If the path contains spaces, it should be surrounded with quotation marks ("). This way, when you run the shortcut, cmd.exe will execute your script and leave its window open so you can work with it. Without the /k parameter the cmd.exe shell would run the script and exit immediately.

Makefiles are usually long and somewhat complicated files. Fortunately you don't have to learn their syntax and write them by yourself. WinAVR contains a nice tool for generating Makefiles, called MFile. This program is actually a Tcl script executed by the so called Windowing Shell (wish), which is part of the Tcl/Tk programming suite. You can run MFile simply from WinAVR group in Windows start menu. It starts with template Makefile already loaded. Now, with *Makefile* menu you can change various compilation settings. This is essentially the same as changing project settings in integrated development environments like AVR Studio. When you make a change, Makefile is scrolled and appropriate entry is highlighted. If you want to edit a Makefile manually, you have to enable editing. Usually it's just enough to set main file name (without extension) and type of microcontroller your source code is written for (*MCU type*). If your project consists of more than one source file (.c file, not .h files) then you need to provide their names (*C/C++ source files(s)…*).

Makefile can be used not only for compilation but for programming too. With MFile you can configure AVRDUDE: set type of programmer and port.

In Makefiles there is a concept of targets, things to be done by *make* command while processing a Makefile file. In default Makefile generated by MFile there are several targets. The default one, executed when *make* command is issued without any particular target specified, is called *all*. It is responsible for compiling source code and generating executable file. It also creates other files, for example HEX file for programmer. If you run the *all* target when files from previous compilation already exist, *make* will check if source code files were modified since last compilation. If there were not changed, it won't do the compilation for them. For easy removal of files created during compilation there is the *clean* target. Thus *make clean* removes compiled files. In order to program a microcontroller, you can use the *program* target which launches AVRDUDE. It requires the code to be compiled (it needs HEX files) so if there are no compiled files, *make* will process the *all* target first. In other words *program* depends on *all*. There are also other targets placed in default Makefile but they are not important now.

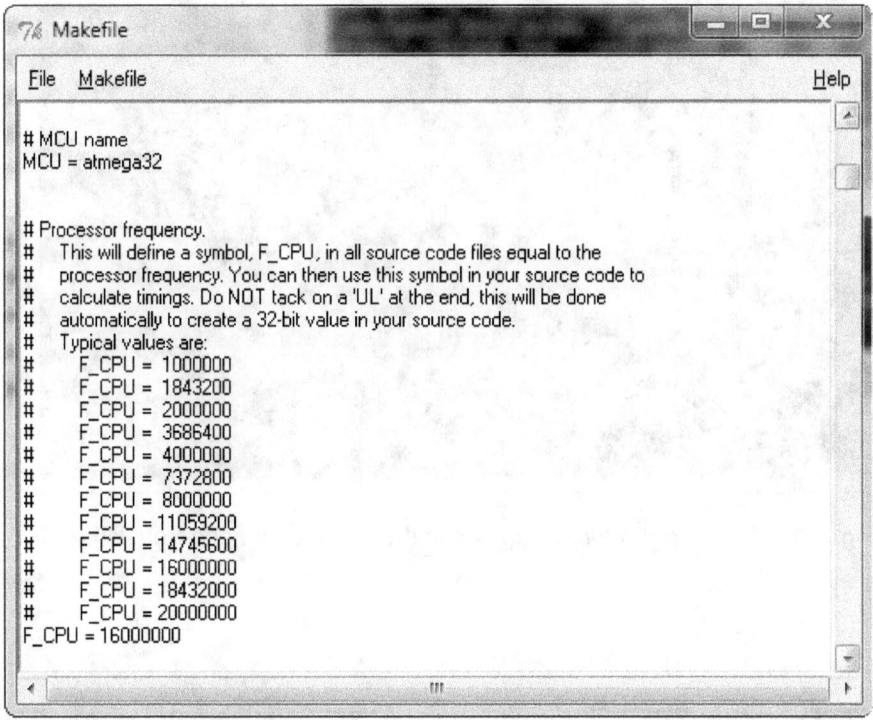

Fig. 3.1.1.1. The MFile editor

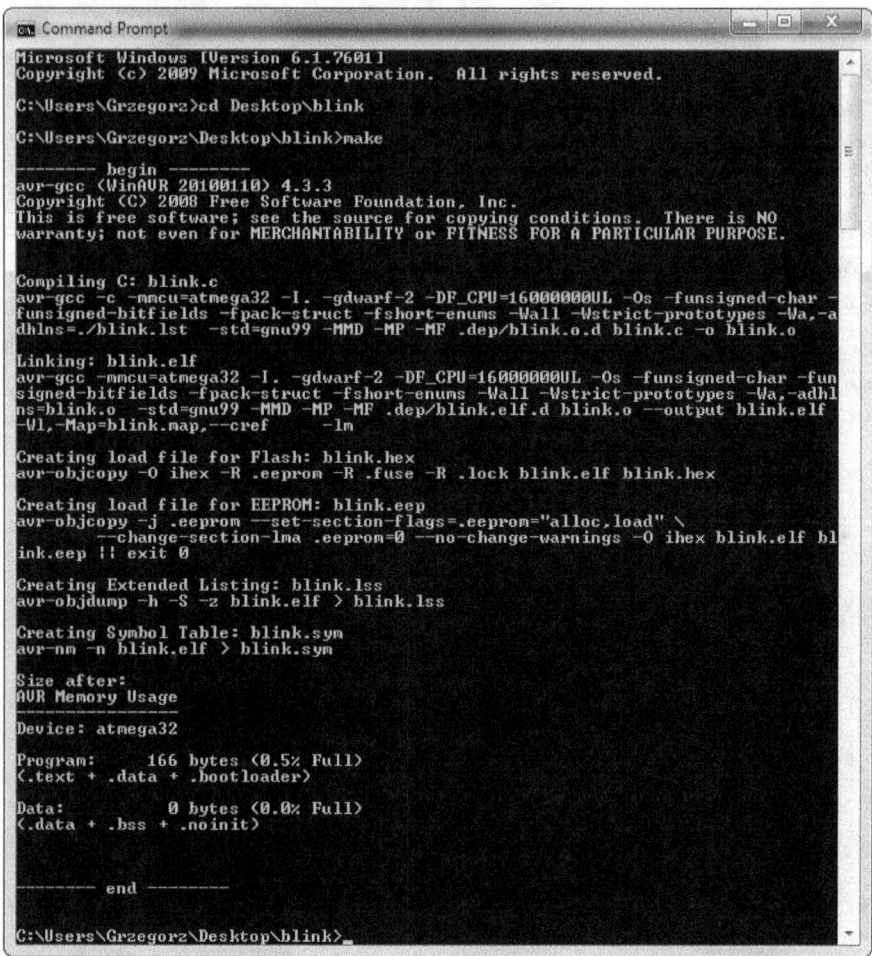

Fig. 3.1.1.2. Compiling a program using the make command from WinAVR

3.1.2. Atmel Studio

Atmel Studio is an IDE delivered by Atmel, available in version 6 at the time of writing this book. It's free but the license forbids creating software for products of Atmel's competitors. In the past (when the product was named AVR Studio 4) it didn't include a compiler and it required WinAVR to be installed. With version 5 the IDE has been completely revamped. It has got its own complete toolchain, based on AVR-GCC. There has been also a new look and feel thanks to Microsoft Visual Studio Shell. Current version is called Atmel Studio 6 because it integrates development tools not only for AVR but also Atmel ARM Cortex-M microcontrollers.

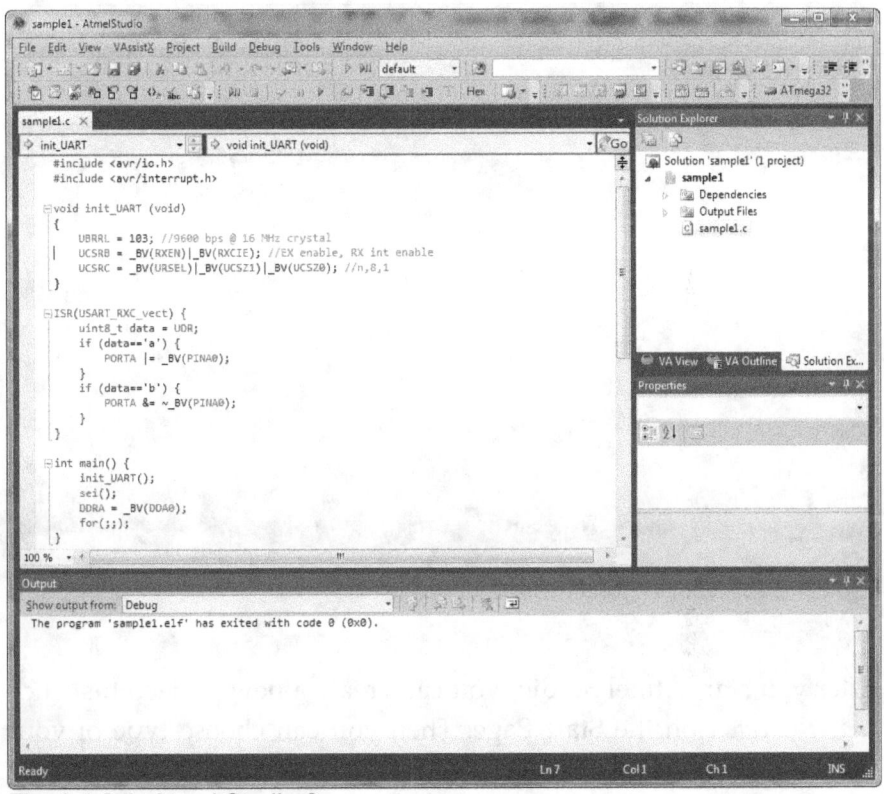

Fig. 3.1.2.1. Atmel Studio 6

You have to know that with AVR Studio 5 Atmel dropped support for older JTAG interfaces, which were seen in Windows as virtual COM ports. Such JTAG is for example JTAG ICE. For AVR Studio 5 or Atmel Studio 6 you need compatible device: AVR Dragon, AVRISP mkII, AVR ONE!, JTAG ICE 3, JTAG ICE mkII, STK500 or STK600.

Serial port and microcontrollers

Atmel Studio installer contains drivers for Atmel USB debugging tools, e.g. AVR Dragon. Don't forget to install them. When Atmel Studio is installed run it and then go to *Tools* menu and select *AVR Tools Firmware Upgrade* to upgrade your Atmel tool if you have one.

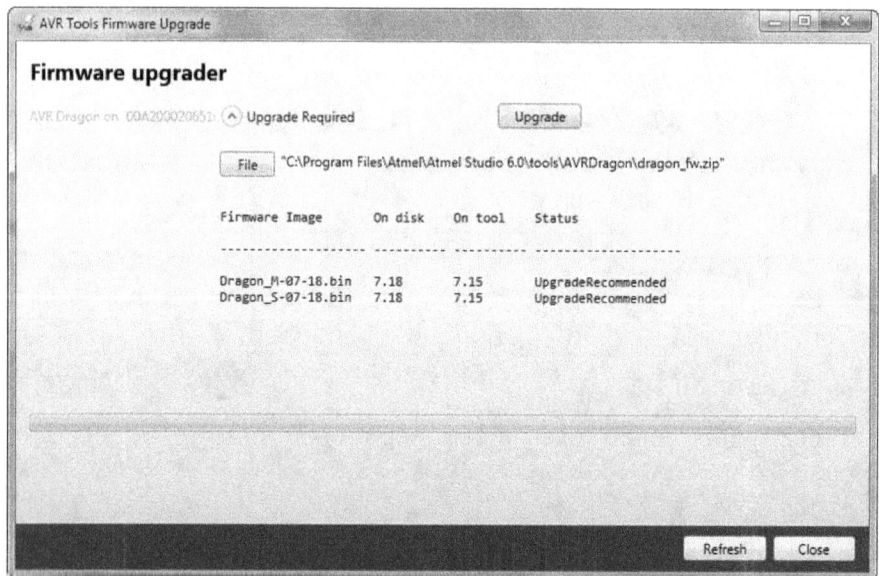

Fig. 3.1.2.2. AVR Dragon firmware upgrade

After you start Atmel Studio, you can create a new project. Just click *New Project...* on the Start Page. Then you can choose type of your project, usually you are going to select *AVRGCC C Executable* Project. At the bottom of the window type name of your project. To have separate folder created for the project, check the *Create directory for solution* checkbox. The next step is selection of the microcontroller you are going to use, for example ATmega32. After you select the chip the project is created and you can start writing code. To select such options like optimization level or debugging device, go to project's properties. You can select there also your debugging/programming device.

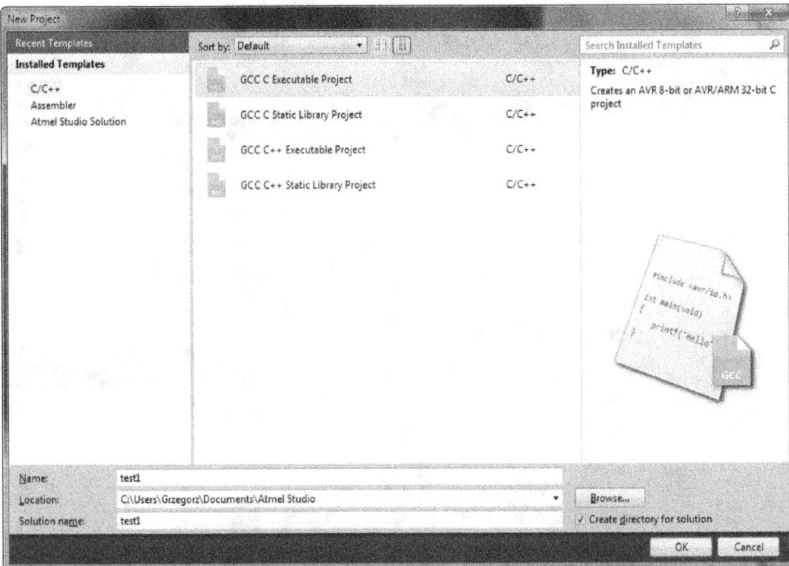

Fig. 3.1.2.3. Creating a new project

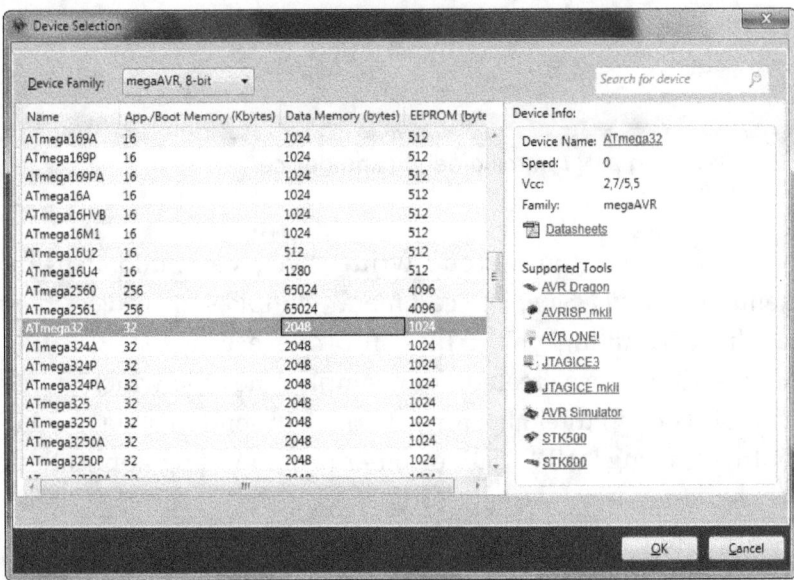

Fig. 3.1.2.4. Selecting target microcontroller

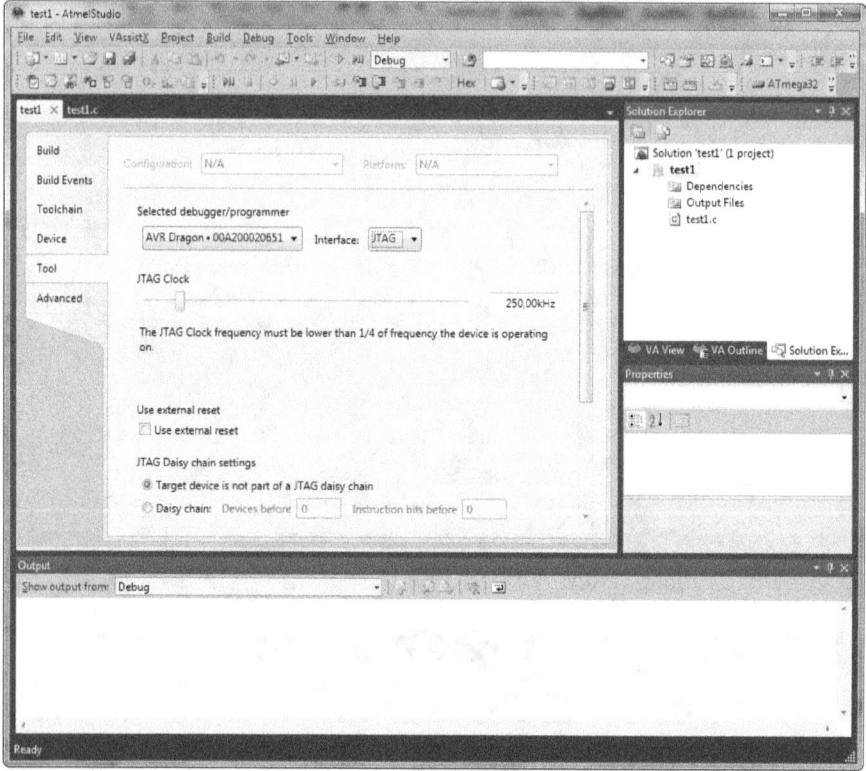

Fig. 3.1.2.5. Selecting programming/debugging device.

When you have your source code written, you can build it using *Build* menu. The *Build Solution* command compiles all project files and then links resulting object files into one executable ELF file. There are also other files created, for example a hex file which stores the executable code in Intel Hex format. You can send this file to your microcontroller using AVRDUDE or other programming application. You can also do it from Atmel Studio if you have a programming device supported by the application. If the device supports debugging then the program is sent automatically when you launch functions such as *Continue* or *Start Debugging and Break* from the Debug menu. If debugging is not supported, you can use *Start Without Debugging*.

Atmel Studio supports debugging which means that you can run your program in step mode and observe state of registers, flags, ports, etc. You can even modify their values. That's very neat functionality. You must know however, that it has limitations. Sometimes debugging can work improperly. It depends what library functions you use, what is the optimization level and which JTAG interface you use. Optimization level can be changed in project configuration. To disable optimization, set it to –O0.

3.1.3. BASCOM-AVR

BASCOM-AVR is a complete development environment for AVR microcontrollers. It differs from other solutions by usage of MCS Basic, a dialect of BASIC programming language. As you can guess, programming in Bascom AVR is usually simpler and faster than writing in C. It comes with built-in functions for RS232, 1Wire, SPI, and I²C protocols and for controlling peripherals such as LCDs or RC5 IR receivers too. Programming in MCS Basic is done at a bit higher level than in C, programmer doesn't need to know so many details about internals of a microcontroller. The downside is of course less control but on the other hand MCS Basic is enough for many applications. Moreover it allows to insert fragments of code in assembly language so you can do virtually everything you want. Bascom AVR supports mostly ISP programmers and has no support for debugging through JTAG interface. It has just a simulator. BASCOM-AVR is a commercial tool developed by MCS Electronics. You can download free version which has limited code size to 4 kB. It means that your compiled, native code can't be larger than 4 kB. When you look at programs running at your PC, which require many megabytes on your hard drive, you can think about 4kB as a very small amount of code. You will see that in microcontrollers' world it's enough for many applications.

Serial port and microcontrollers

The interface of BASCOM-AVR is quite simple but has some useful features too. The *Program* menu offers basic commands for compilation, syntax checking, and programming. From this menu you can also reset you microcontroller or start simulation. Programming can be done automatically or you may open a programmer window. It is very similar to stand alone software programmers, e.g. ISP Programmer by Adam Dybkowski. It allows to read, write and verify chip's Flash memory. It also gives access to EEPROM and fuse bits. Simulator is a tools using which you can test your program without launching it on actual hardware. You can run your program step by step and observe contents of registers, I/O registers, memory, and flags.

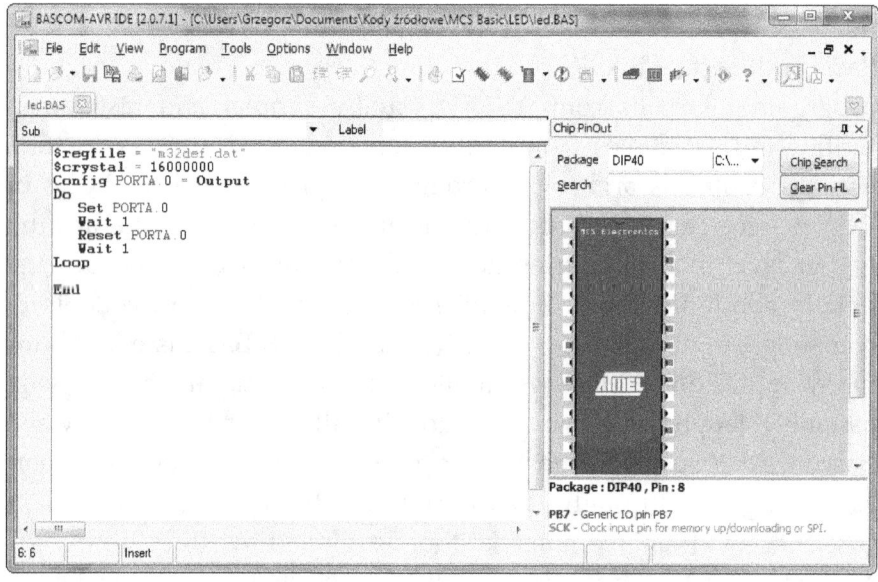

Fig. 3.1.3.1. BASCOM-AVR

The *Tools* menu gives access to a few interesting features. One of them is *PDF Update* which offers and easy way to download data sheets for AVR microcontrollers. Once one or more of them are downloaded, you can later check if new versions are available. BASCOM-AVR comes even with internal PDF viewer which can be launched using *View* menu. The other nice tool is *Easy TCP/IP* which can be used to test network connectivity. It allows you to run simple TCP server and connect to it using one or two internal clients. Of course you can make more connections using other applications or another computer(s). The situation is similar for UDP but the interface is different because of the fact that UDP is a connectionless protocol. *Easy TCP/IP* has also integrated simple Base64 coder/decoder. BASCOM-AVR offers two useful tools for LCDs. One is *LCD designer* which helps to create a command for defining custom character for alpha-numeric LCDs. The second one is *Graphic Converter* which is intended to convert BMP files into BASCOM Graphic Files (.BGF) that can be used with graphic LCD displays.

3.1.4. Eclipse

Eclipse is an Open Source development environment used for programming in many languages. It is often used for writing Java software but also can be used for programming in C for AVR and ARM microcontrollers. Thus if you have some experience with Eclipse you may consider using it for AVR programming. Eclipse requires Java Runtime Environment (JRE) or Java Development Kit (JDK) installed. Moreover the application exists in several versions, you need to download the one designed for C/C++ development.

Eclipse itself, just after installation, isn't ready yet for developing software for AVRs. First of all it needs WinAVR installed. Secondly it must be properly configured in order to utilize WinAVR tools. Fortunately there is no need to do the configuration manually. There is an appropriate plugin available at http://avr-eclipse.sourceforge.net/ But you don't even have to open this site although it's good to take a look at it. All you really need is to click *Help* menu and select *Install New Software…* In the *Install* window type http://avr-eclipse.sourceforge.net/updatesite/ into *Work with:* field. After a short moment you will see *CDT Optional Features* on the list below. Expand this item, select *AVR Eclipse Plugin*, and click *Next*. Then you need to click *Next* again and accept the plugin's license terms. The plugin will be automatically downloaded and installed. During installation you may get a warning about unsigned software, then just confirm installation.

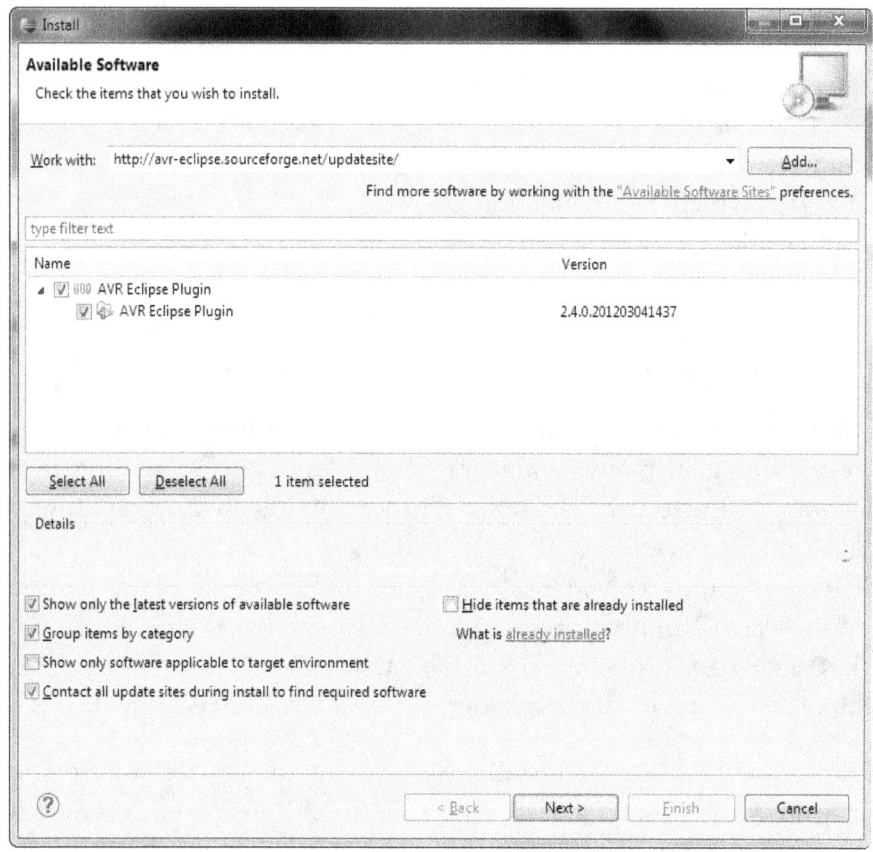

Fig. 3.1.4.1. AVR plugin installation

To create a new project, click *File*, *New* and *Project...* From the *Wizards* tree select *C/C++* and then *C Project*. Now Eclipse will try to run avr-gcc.exe. If WinAVR is not properly installed, an error window will be displayed. If everything is OK, you can type a name for your project. As a type select *AVR Cross Target Application/Empty*

Project. Click *Next* to open next screen when you can select target configurations. You can leave both *Debug* and *Release* checked. The last screen of the wizard allows you to choose type of AVR microcontroller and frequency of its clock. At this moment the project is created but it's empty. You can add a source file using *File* menu and selecting *New* and *Source File.* When you are finished with writing source code you can compile it by clicking *Build Project* from the *Project* menu.

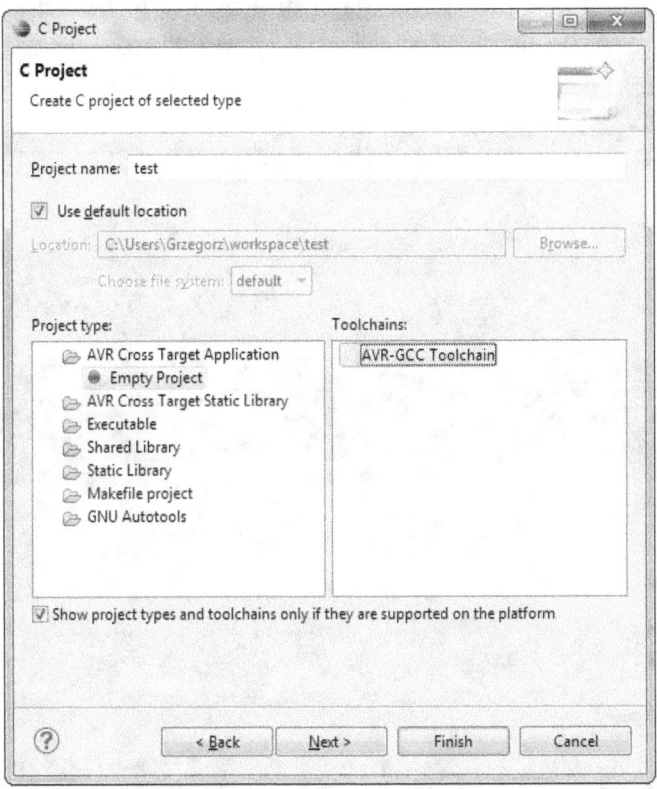

Fig. 3.1.4.2. New project cration

If you wish to change settings such as microcontroller type, clock frequency, or optimization then from *Project* menu select *Settings.* In settings window open *C/C++ Build* section and select *Settings.* Here you can change parameters of a given configuration (Debug, Release or other). Eclipse with WinAVR plugin can also be used for programming a chip using AVRDUDE. To configure this program select *AVR* and then *AVRDude*. Here you can select and configure you hardware programmer, e.g. STK200. When you are done and you programmer is connected, you can try clicking *AVR* menu and selecting *Upload Project to Target Device*. Most probably you will get an error saying there is no HEX file. In this case go to build settings and select *Additional Tools in Toolchain*. Here you can enable generating a HEX file(s) for Flash and/or EEPROM which can be loaded by AVRDUDE. To actually create these files build your project.

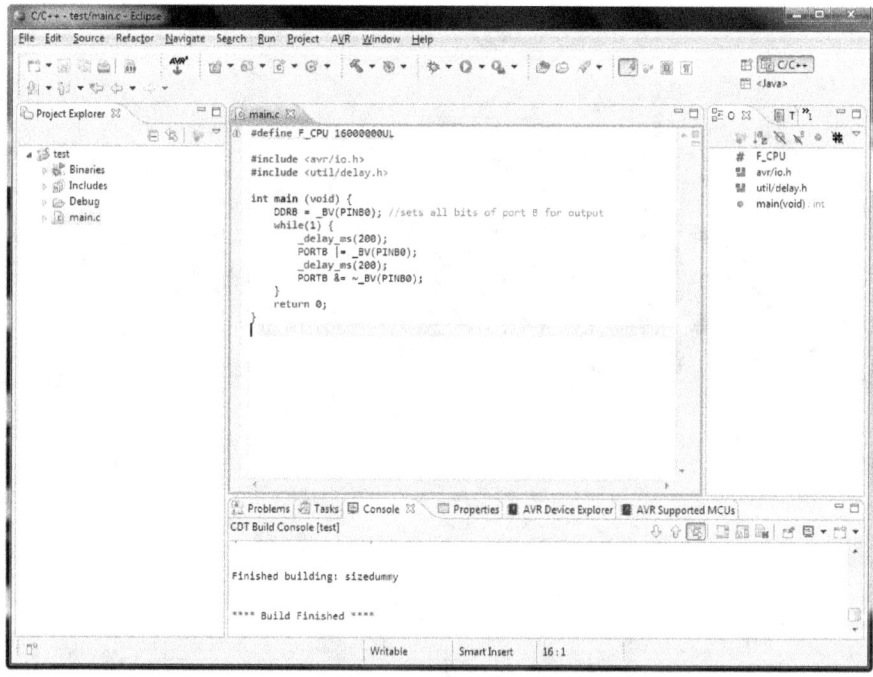

Fig. 3.1.4.3. Main Eclipse window

3.1.5. ISP Programmer by Adam Dybkowski

It's a tiny, freeware application for programming AVR microcontrollers through ISP interface, written by Adam Dybkowski. It has simple GUI which allows to program microcontroller's Flash and EEPROM memory and set fuse bits. It can be used to program a microcontroller using .hex files generated by AVR GCC. ISP Programmer application supports popular hardware ISP programmers, including STK200. It is available at http://dybkowski.net/content/en/node/15

The program is easy to use. After first launch, click the Setup button and select your programmer. There is a list of popular programmer. If the one you own has non-standard pinout, configure pins manually. Inverted RESET option applies to some non-standard programmers. It doesn't have any connection to the fact that reset signal is negated in AVRs (activated by logical zero). In some applications the reset signal is generated by specialized circuit and the programmer has to provide logical one as external reset signal for ISP programming. Normally this option should be left unchecked.

When the setup is complete the last thing to configure is speed of programming. The higher clock rate you select, the shorter time programming will take. However the speed can't be too high. For most AVRs the clocking of SPI transmission must be at least 4 times lower than clocking of a microcontroller. In the top-right corner of application's window select clock rate of your chip. The application will calculate appropriate speed basing on type of the microcontroller. The FASTEST setting causes ISP Programmer at maximum speed of used parallel port.

When the programmer is connected to AVR microcontroller and to a PC and the power for the microcontroller is turned on, you can test the whole setup. Just click Read signature button. The application should display model of your AVR microcontroller, including size of

Flash and EEPROM memories. If it works you can perform actual programming.

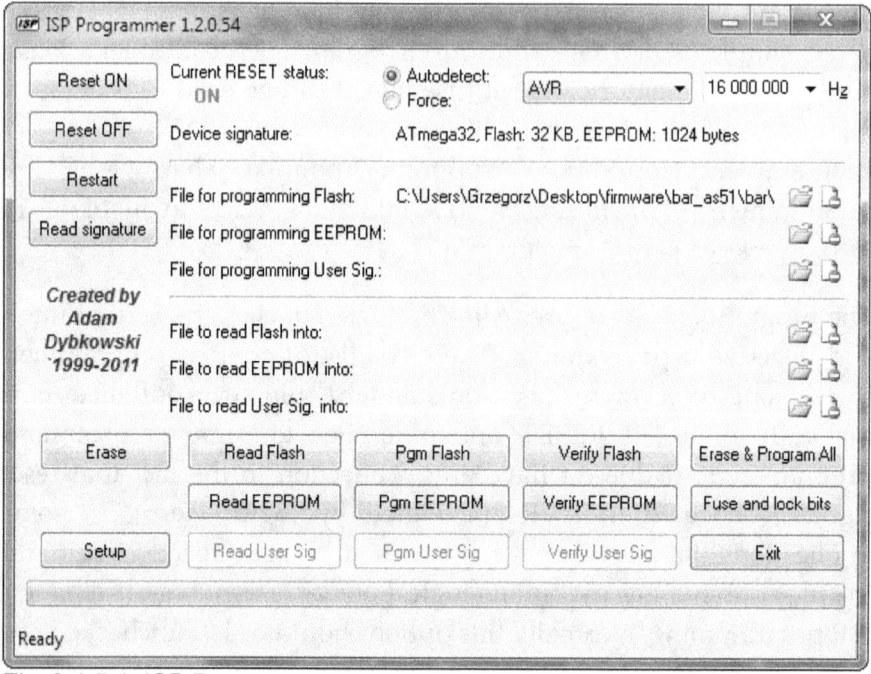

Fig. 3.1.5.1. ISP Programmer

There are several different things in a microcontroller you can program: Flash memory which stores executable code, EEPROM which is non-volatile storage for data, user signature and fuse and lock bits. Fuse bits are used to configure a microcontroller and they depend on the type of a chip. For example if a microcontroller supports JTAG interface, there is JTAGEN fuse bit which allows us to enable or disable the interface. There are CKSELx fuse bit which are responsible for selecting clocking options. For ATMega32 you can use 5 clock sources: external crystal/ceramic resonator, external low-frequency crystal, external RC oscillator, calibrated internal RC oscillator, and external clock. By default internal RC oscillator is used. If you want to use another source of clocking signal you have to set appropriate bits. You should be very careful when changing fuse bits. If you program the bits in wrong way a microcontroller may become inaccessible. The common mistake is selecting external clock instead of crystal resonator. In such case you have to connect appropriate square wave generator to make a microcontroller working. Mistakes are usually an effect of one tricky fact. In datasheets of AVRs you can find fuse bits described as programmed or unprogrammed. You must remember that programmed fuse bit has value 0 and unprogrammed one has value 1. This can be confusing so pay attention. In ISP Programmer application numerical values are used so possibility of a mistake is rather small.

Programming Flash and EEPROM is easy. You just have to select appropriate Intel-HEX file. In case of some AVRs you can also program user signature. If a microcontroller is already programmed you have to delete contents of its memory by clicking Erase. You can verify correctness of programming using Verify (…) button. This performs the programming and then reads the written data and compares it with the file used for programming.

When reset signal is turner off, microcontroller starts to execute programmed code.

3.1.6. PonyProg2000

PonyProg2000 (or just PonyProg) is a free, open source application available for both Windows and Linux platforms. It supports two types of programmers: serial (SI Prog, JDM) and parallel (AVR ISP, DT-006, EasyI2C). The SI Prog programmer was designed by the author of PonyProg2000. It is described in more details in one of following chapters. AVR ISP refers to popular STK200 programmer which is also covered in this book. At first run PonyProg2000 asks for calibration which can be performed using Setup menu. When program is calibrated, there are two more things to set. First one is a type of IC to program. The application supports many chips, not only AVR microcontrollers. You can select you chip using Device menu or

Serial port and microcontrollers

drop-down lists. The second thing we need to do is selection of hardware interface. Most programmers appear two times: wit API and I/O suffixes. Choose API on Linux and I/O on Windows. When the programmer is selected, choose appropriate port. You can test the configuration by clicking Probe button.

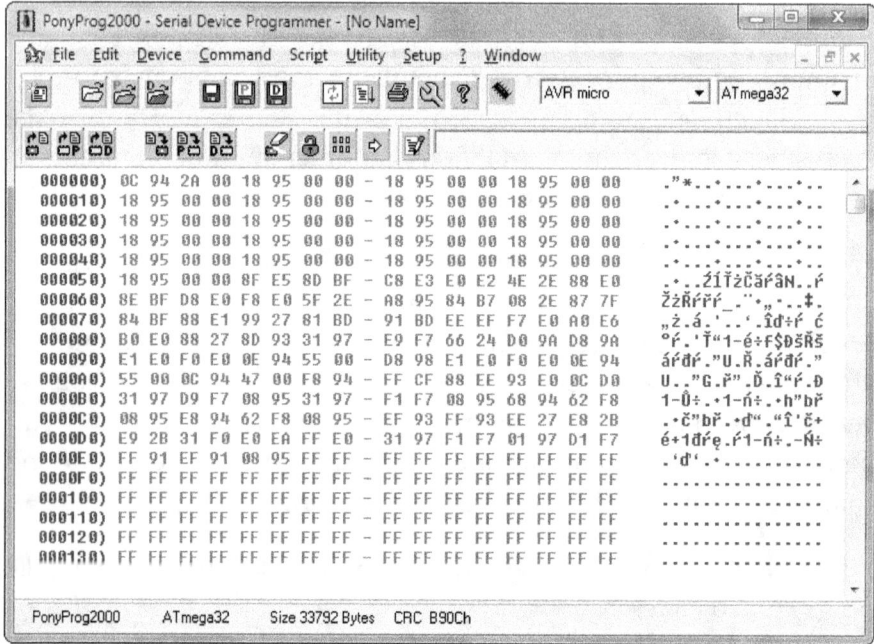

Fig. 3.1.6.1. PonyProg2000

When you are done with configuration, you can work with a microcontroller. The Command menu gives access to functions such as programming, reading and verifying contents of Flash and EEPROM memory and also setting fuse bits. To program a microcontroller, first open appropriate HEX file by clicking File menu and selecting Open Program (FLASH) File... Don't forget to select *.hex file type in *Open program (FLASH) content file* window when navigating to appropriate directory. After you loaded the HEX file you can send it to your microcontroller. PonyProg2000 will perform automatic verification by reading the memory back.

An interesting feature the program is support for scripting. You can write own scripts (E2S files) to automate frequent tasks. We won't be discussing this topic here, you can find necessary information in PonyProg2000 documentation.

If you start the program and get driver error run it with administrative privileges.

3.1.7. AVRDUDE

The program known as AVRDUDE was originally named avrprog and was available for FreeBSD operating system. Later it was renamed to AVRDUDE and released for other platforms, including Linux and Windows. It supports large variety of microcontrollers and programmers. AVRDUDE itself is a console program, working in text mode (including interactive mode). However there are also graphical interfaces available. Because AVRDUDE can be configured with command line parameters it can be used as an external programmer for such applications as BASCOM AVR or Atmel Studio. Thanks to AVRDUDE you can use hardware programmers which are unsupported by these development environments.

AVRDUDE is very easy to install under Linux. On Ubuntu it can be installed using standard apt-get command:

```
apt-get install avrdude
```

Of course we need sudo if we don't have root privileges. Not only installation requires such privileges, running AVRDUDE requires root too. Because root account is not intended for normal work but for administrative tasks, we should run AVRDUDE from normal account. To do this, still being logged as root, we need to set SUID bit for program's binary. It can be done with following command:

```
chmod u+s `which avrdude`
```

Serial port and microcontrollers

The which command finds location of the binary and returns it to chmod program which sets SUID bit for all owner of executable. The binary is owned by root so it will be executed also as root.

Now, when the permissions are configured, we can check if everything is working. Of course we need a programmer and microcontroller connected and powered up. Let's enter a command like this:

```
avrdude -c stk200 -p m32
```

The -c parameter specifies a programmer and -p parameter specifies a particular AVR microcontroller. In this example we have STK200 programmer and ATmega32 chip. A list of supported hardware and AVRs is located in /etc/avrdude.conf. If we didn't make any mistake, AVRDUDE should read microcontroller's signature. For ATmega32 it's 0x1e9502. How can we know if the signature was read correctly? We can check it in our chip's data sheet. For ATmega32, Atmel writes in Signature Bytes section:

```
For the ATmega32 the signature bytes are:
1. $000: $1E (indicates manufactured by Atmel)
2. $001: $95 (indicates 32KB Flash memory)
3. $002: $02 (indicates ATmega32 device when $001 is $95)
```

So everything matches.

To erase a microcontroller's memory we use -e parameter, e.g.:

```
avrdude -c stk200 -p m32 -e
```

To perform other operations on contents of memory we need -U parameter. It must be supplemented with additional parameters: type of memory, type of operation, and file name. Usually we will operate on two types of memory EEPROM and Flash. AVRDUDE allows as well manipulating contents of other areas and types of memory: fuse bits, boot area, lock byte, or user signature. As for type of operation there are three: write (w), read (r), and verify (v).

Having this information in mind we can write a command for programming our microcontroller:

```
avrdude -c stk200 -p m32 -U flash:w:program.hex
```

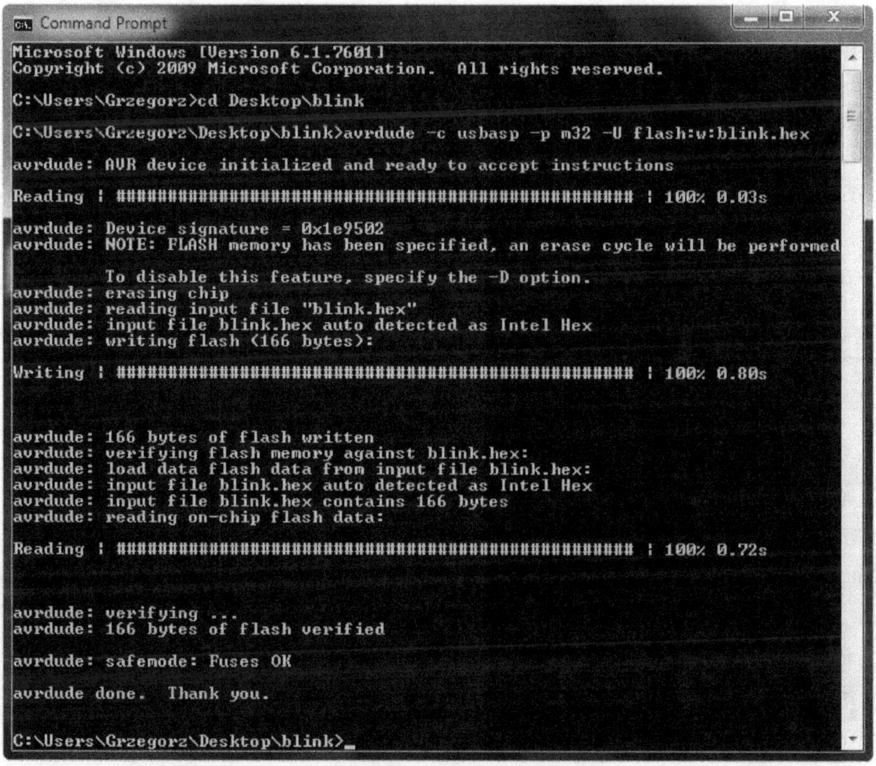

Fig. 3.1.7.1. AVRDUDE

If we change w to r then the contents of Flash memory would be read into program.hex file. If we use v, than contents of chip's memory will be compared against program.hex.

For writing fuse bits we use immediate values (m):

```
avrdude -c usbasp -p m8 -U hfuse:w:0xC9:m -U lfuse:w:0x9F:m
```

Reading is done like for Flash, using files:

Serial port and microcontrollers

```
avrdude -c usbasp -p m8 -U hfuse:r:high.txt:h -U
lfuse:r:low.txt:h
```

The range of available functionality depends on a microcontroller and a programmer we use. For example using

```
-E reset
```

parameter we can leave a microcontroller in reset state after programming. However this feature won't work with some programmers, e.g. standard version of STK200. After programming the tri-state buffers in this programmer are put in high impedance state, and even if appropriate line of parallel port is set to low state by AVRDUDE, the microcontroller's RESET pin won't be set to low state (RESET pin in AVRs is negated so reset condition is triggered by low logical state).

AVRDUDE for Windows comes with WinAVR package. You can find avrdude.exe in the bin subdirectory, where it is located with other applications, such as the avr-gcc.exe compiler. WinAVR provides also the giveio driver which allows AVRDUDE to access ports on Windows. If you try to run AVRDUDE on Windows and see some error messages related to giveio make sure it is properly installed. Status of giveio may be checked with following command (must be launched with administrative privileges):

```
loaddrv status giveio
```

Usage of Windows version of AVRDUDE is the same as Linux version, you don't need to learn new commands. Just open the command prompt (cmd.exe) navigate to appropriate directory (by default it's *C:\Program Files\WinAVR\bin*) using cd command and run avrdude.exe with parameters you need. You can also add the directory to the environmental variable PATH so you can run AVRDUDE from any directory, for example the one you use store you source code in. This will also allow you to run avr-gcc.exe from any directory.

78

The most popular graphical interface for AVRDUDE is AVR8 Burn-O-Mat. It is written in Java and can be run on both Windows and Linux, it just needs Java Runtime Environment (JRE) installed. The program has a form of a JAR archive and is executed by Java Virtual Machine. The JAR file comes with two scripts which can be used to launch it. There a two one-liners, one for Windows (.CMD), the second one for Linux (.SH). If the JAR file is properly associated it can be also run by a double-click, just like any other application. If you have trouble launching AVR8 Burn-O-Mat, first of all check if your JRE version is not too old, an upgrade may be necessary.

After first run, simple configuration is needed. You have to provide locations of AVRDUDE binary and configuration files. You also need to choose you programmer and port it uses (e.g. lpt1). Select you microcontroller in main window as well. When AVR8 Burn-O-Mat is configured you can start using it, which is very easy. You can write, read, and verify Flash and EEPROM memory. You can also set fuse bits. When a checkbox next to bit's name is checked it makes it programmed (value set to 0). The very nice thing about AVR8 Burn-O-Mat is that it can help you with settings fuse bits. Normally you would have to carefully study the datasheet of you microcontroller, learn meaning of each fuse bit and what the dependencies between them are. With this program however, you can just set desired options and appropriate fuse bits will be set. There is *Clock Options* button which opens a window where you can set all the options related to clocking signal. The second button, *Misc Options*, opens a window with other settings. By default fuses window operates in normal mode which means you can't edit some fuse bits. It is possible after you switch to expert mode.

AVR8 Burn-O-Mat is not the only GUI for AVRDUDE you can find on the Internet. For example there is an application called avrdude gui v.2.0. It has one window containing whole configuration. There is only one button, Execute, which is used to run AVRDUDE. It depends on checkboxes what actions will be performed. It makes the program complicated. Moreover it requires fuse bits to be written in hexadecimal values, not as single bits. This means that you have to make necessary calculations. *avrdude gui* allows also to easily check if giveio works properly, you just need to click Status button. This function requires the application to be run with administrative privileges.

3.2. *Hardware tools*

When we have our code written and compiled we have to put is somehow into microcontroller's flash memory. It is usually done in two ways: by ISP or JTAG interfaces. Each one of them requires different hardware devices which are connected between a PC and the programmed microcontroller.

3.2.1. ISP programmers

The most popular way of programming AVR microcontrollers is with ISP programmers which use SPI protocol. They are connected to parallel port or USB. Actually one doesn't need any hardware programmer if a parallel port is available. If a microcontroller is powered from 5V power supply it can be directly connected to parallel port using just wires. This approach is not very common though. Parallel port is almost always integrated with the motherboard and damage to the port would mean necessity of replacing the motherboard or buying external parallel port PCI card. Thus we want to have some kind of a buffer between a PC and programmed AVR microcontroller. The most popular such device is STK200. It is very easy and inexpensive to build. It consists of one chip – 74HC244. There are also three additional elements: filtering capacitor, pull-up resistor and a diode. All those elements are easily accessible and available at very low cost. You can mount them on very simple PCB or even solder them directly to DB25M connector. If you want you can of course buy assembled STK200 programmer, for example ZL2PRG from Kamami.

Serial port and microcontrollers

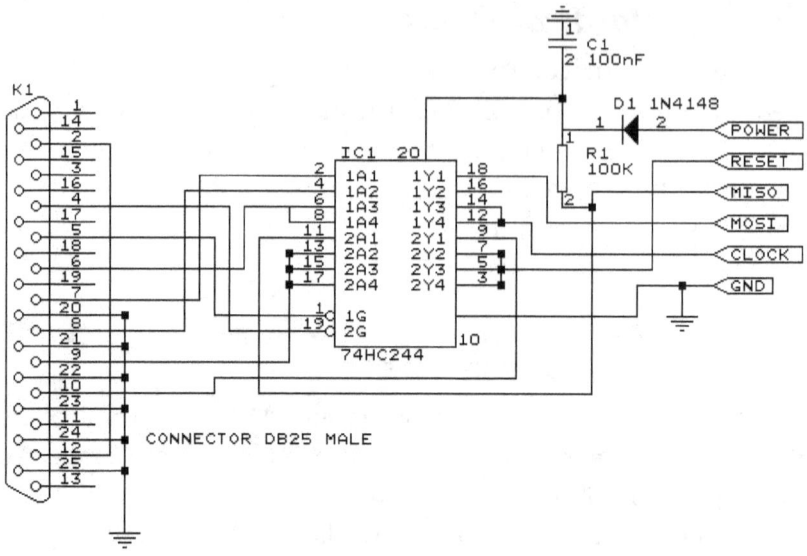

Fig. 3.2.1.1. Schematic of STK200 programmer

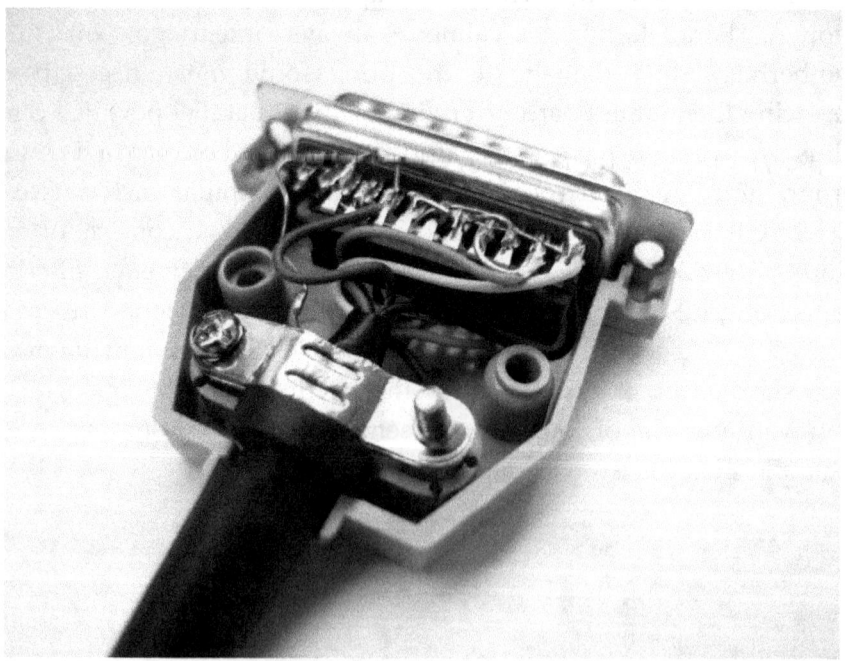

Fig. 3.2.1.2. STK200 mounted inside DB-25M plug

If you don't have a parallel port, which is very likely nowadays, you have two choices: add PCI or ExpressCard expansion card to your computer or use a USB programmer. Don't try parallel-USB adapters, they are made for printers and may not work correctly. One of the most popular USB programmers is USBasp. It's easy to build but as it's based on AVR itself you need another programmer to program the microcontroller. You can also buy a USBasp kit with preprogrammed chip. USBasp is supported by AVRDUDE, eXtreme Burner – AVR, and BASCOM AVR. USBasp can be used on Windows, Linux and Mac OS X. On Windows installation of a driver is required. All information about USBasp you can find on official site: http://www.fischl.de/usbasp/ There are also presented various versions of this programmer built by people. An interesting one is by Paweł Szramowski who added a buffer which allows to program microcontrollers powered by 3.3V supply voltage. USBasp is very easy to use. The only thing you have to keep in mind is *Slow SCK* jumper. It has to be present if programmed microcontroller is configured with slow clock, like 1 MHz which is often the default.

Fig. 3.2.1.3. USBasp, low voltage version

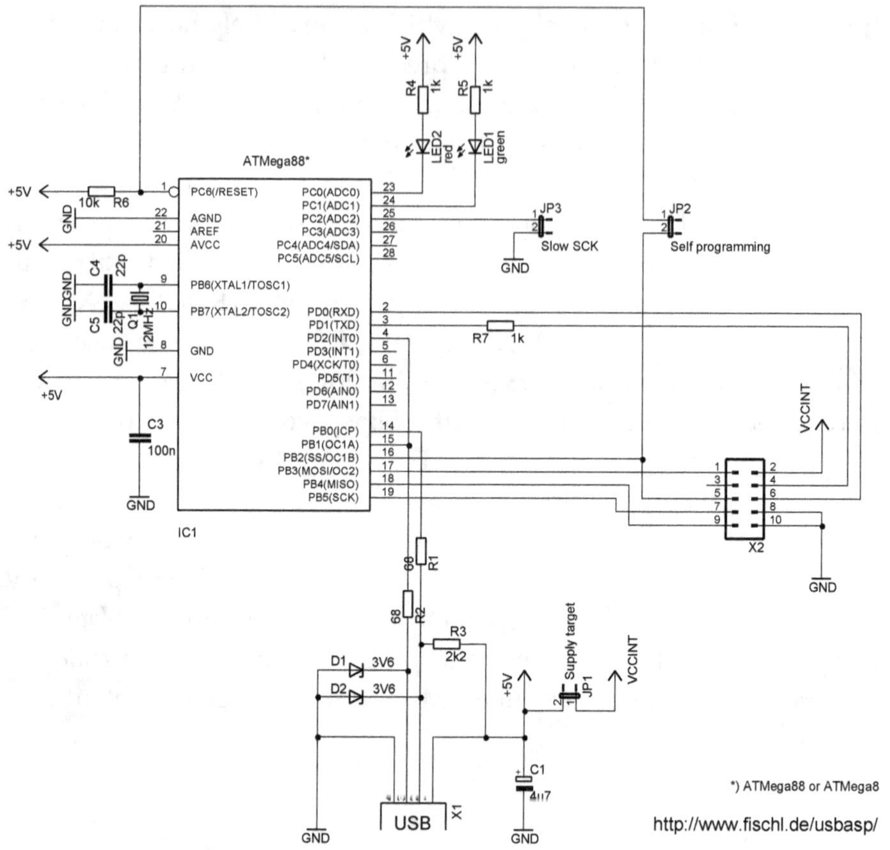

Serial port and microcontrollers

Fig. 3.2.1.4. Original USBasp circuit

http://www.fischl.de/usbasp/

*) ATMega88 or ATMega8

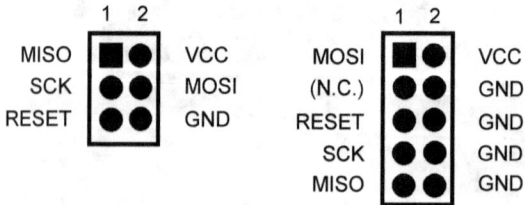

Fig. 3.2.1.5. ISP connectors: 6-pin and 10-pin

There are two standard connectors for ISP programmers: 6-pin and 10-pin headers. The STK200 programmer uses the 10-pin version, the ZL15AVR development board uses it as well. If you buy a programmer and a development board make sure, that they use the same socket convention. If you design your own PCB, you have to make the decision basing on the programmer you already have or intend to obtain. Both connectors have the same signals. In the 10-pin connector the 4 extra pins are connected to ground. This gives total 5 or 6 pins connected to ground in this type of ISP connector. Actual number depends on the role of pin 3. Sometimes this pin is not connected to anything, sometimes it's connected to ground. If you are creating own PCB with ISP connector or own ISP programmer and you are not sure, just leave this pin unconnected. In case of 6-pin connector there is no problem because all pins are used: two for power supply and four for actual programming.

3.2.2. JTAG interfaces

JTAG is actually an acronym for Joint Test Action Group which published Standard Test Access Port and Boundary-Scan Architecture. This standard was designed for testing printed circuit boards using boundary scan but it found also other applications: debugging and programming integrated circuits. JTAG is commonly used as a name for devices used for these two purposes. They have usually two connectors: USB for connecting to a computer and 10- or 20-pin JTAG interface for connecting a microcontroller. One of JTAGs offered by the Kamami company is known as ZL16PRG. It has 10-pin JTAG interface for connecting with ZL15AVR development board. It's a low cost option but is not supported by AVR Studio 5 and newer versions. Therefore it's good choice for AVR Studio 4 users.

Another JTAG interface available from Kamami and other companies as well is AVR Dragon, produced by Atmel. Aside of JTAG interface it also supports ISP (In-System Programming), HVSP (High Voltage

Serial Programming) and PP (Parallel Programming). As the last two are not very popular there are no connectors available for them on the board, you need to mount pins by yourself if you intend to use those interfaces.

You can connect AVR Dragon to ZL15AVR with standard AVR JTAG cable with no problem, just pay attention to location of pin 1. As ZL15AVR offers JTAG, there is no need to program it using ISP if you have AVR Dragon. ISP is just for programming while JTAG additionally allows debugging of programs. If you want to use ISP anyway, you need appropriate cable because AVR Dragon has 6-pin connector and ZL15AVR has 10-pin connector.

Fig. 3.2.2.1. AVR Dragon programmer

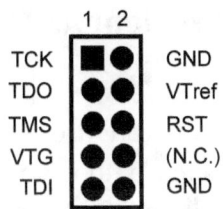

Fig. 3.2.2.3. JTAG connector

Fig. 3.2.2.2. ZL16PRG JTAG interface

3.2.3. ZL15AVR development board

For easy and quick start you can use a development board as a hardware platform. With such a tool you can start programming with a little less detailed knowledge about hardware. Moreover development boards are equipped with various circuits which you can connect to the microcontroller. These are usually LEDs, buttons, memories, LCDs, real time clocks, voltage converters, buzzers, transistors, etc. From a circuit diagram from board's documentation you can tell what are the connections on a board and how to connect specific device to the microcontroller.

Examples in this book are based on ZL15AVR development board which is equipped with ATMEGA32 microcontroller. This chip has 32 kB of flash memory, 2 kB of SRAM and 1 kB of EEPROM. It has two 8-bit timers/counters, one 16-bit timer/counter, real time counter, four PWM channels, 8-channel 10-bit ADC, I²C interface, USART, SPI interface, watchdog and analog comparator. JTAG interface provides an easy way for programming and debugging. The ZL15AVR development board is equipped with such peripherals as: M41T00 real-time clock, TC77 digital thermometer, MCP9701 analog thermometer, TSOP31236 infra-red receiver, a piezo buzzer, 8-segment LED display, DB9 connector for RS232, buttons and LEDs.

Of course you can use any development board you like and which is suitable for your needs. You can also build own circuit for your AVR microcontroller.

Fig. 3.2.3.1. ZL15AVR development board

3.3.   *Designing own hardware for AVRs*

If you are building a specific device and you know exactly how it will look like from electrical point of view, you can assemble a PCB and then program the microcontroller. For learning and testing purposes it's better to buy a development board or a breadboard socket. The first is a ready to use piece of hardware, containing essential elements. The second one is just a plain socket where you have to connect everything by yourself. This is of course not easier but gives a lot more experience with designing circuits for microcontrollers.

The breadboard socket (fig. 3.3.4) is a plastic board with contacts designed to make connections without soldering. A circuit can be assembled by simply inserting wires and pins of elements into holes. These holes are connected according to a pattern. When you purchase a breadboard socket you will get a detailed description of these connections. You can make additional connections using wires. With a breadboard you can make your very own circuit, just as you like. You have to insert your microcontroller, all required elements and make necessary connections. This circuit can be just for learning purposes or it can be the circuit you designed for your device. Thus using a breadboard can be a step between using a development board and making target PCB. On the other hand a breadboard can be an inexpensive alternative to a development board and you can use both for learning and testing of your circuits.

In any case at some point you will want to make your own device on a PCB and you will need to know how to connect your microcontroller to the rest of the circuit. Fortunately it's not complicated in case of AVRs. As an example let's take a look at ATmega32, the chip used in ZL15AVR development board. This is one of more powerful AVR microcontrollers and has 40 pins. Fortunately you don't have to bother with all of them.

The essential thing for a microcontroller is that it has to be powered. When we look at ATmega32 datasheet we can identify VCC (10) and GND (11) pins. Thus we have to connect pin 10 to +5V supply voltage and pin 11 to ground. But on pinout diagram we can see some strange pins: AVCC (30) and AGND (31). These are supply voltage pins for Port A and the A/D converter. Even if you don't intend to use port A or the A/D converter, you have to provide supply voltage to them. So connect pin 30 to +5V and pin 31 to ground. If you wish to use A/D converter then you should provide +5V supply voltage to pin 30 through a low pass filter. This will suppress noise present in power supply circuit. If you process some analog signal through A/D converter you wouldn't like to have it distorted.

At this moment the microcontroller is almost ready. After power up it will start to execute program stored in its flash memory. Of course you won't see anything because no output devices are connected. How they should be connected depends on your program – what peripherals you use and what ports you have chosen.

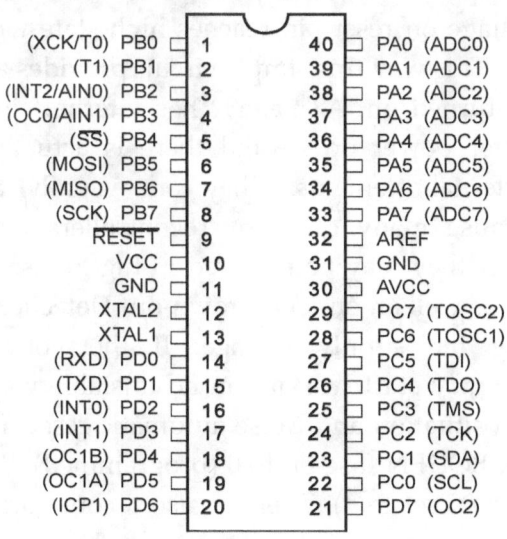

Fig. 3.3.1. Pinout of ATmega32 in PDIP package.

But let's see what else can be connected but is not a typical peripheral circuit. Almost every microprocessor device has a reset button. Sometimes written code is buggy and the program hangs. Sometimes we just want to start an execution over without turning power off and on. ATmega32 has reset input on pin 9. As you can see in a datasheet, it is a negated input which means that the microcontroller is in reset state when there is logical zero on this pin. The pin is internally connected to VCC through a pull-up resistor so it is in high logic state when not connected. So to have a reset button, connect it between pin 9 and ground. When you press the button, it will short-circuit it to the ground and cause low logical level on reset pin pushing a microcontroller into reset state.

If you examine available circuits with ATmega32 or other microcontrollers you can often notify an RC circuit connected to reset pin. This circuit consists of a resistor connecting reset pin to VCC and a capacitor connecting reset pin to GND. When the power supply is turned on, the capacitor is discharged and starts to charge. As it takes some time, voltage on reset pin reaches high state with some delay after power on. This way this simple circuit provides automatic reset after power is turned on. ATmega32 has internal circuit providing reset signal after power on so and there is actually no need for additional, external elements. They are usually added simply habitually because many other microcontrollers need such reset circuit. However they may be required in high noise environments. ATmega32 has as well an on-chip Brown-out Detection (BOD) circuit for monitoring the supply voltage. If the voltage drops the microcontroller may start working in an unstable way. BOD circuit can detect such situation and cause automatic reset. To enable BOD you have to set BODEN fuse bit to 0 (programmed). By default BOD is disabled (BODN set to 1). There is also a configuration bit called BODLEVEL. If it is set to 1 (unprogrammed) then reset occurs when VCC drops to 2.7 V. If you program the bit (set it to 0) then the threshold will be set to 4.0 V. However there is a trick. ATmega32 as many other AVRs is available in both standard and low power version (ATmega32L). Standard ATmega32 can operate at voltage between 4.5 V and 5.5 V while ATmega32L can work at the voltage as low as 2.7 V up to 5.5V. BOD circuit needs appropriate voltage to work so in case of ATmega32 it may be not able to properly reset a microcontroller at 2.7 V. Thus BODLEVEL = 1 is not applicable for ATmega32.

The other thing which can be connected to ATmega32 is a crystal resonator. It can be connected to pins 12 and 13. Additionally each of these pins should have a capacitor connected between the pin and ground. Both should be 12-22pF. But do we need a crystal? ATmega32 has an internal RC oscillator which can work as a clock source for the microcontroller. Being RC circuit, it's not very precise and shouldn't be used in timing sensitive applications. So depending on kind of device you are designing you have to decide whether internal oscillator is enough or not. Typically it has an accuracy of a few percent or worse but can be calibrated to achieve ±1% using OSCCAL register. Depending on CKSEL fuse bits ATmega32's internal oscillator can work at 1, 2, 4 or 8 MHz. For each of these frequencies there are precalculated calibration bytes stored in a microcontroller. They can be read using an ISP programmer. Writing such a precalculated value into an OSCCAL register during runtime assures accuracy of ±3%. To go down to ±1% you have to obtain the calibration value by yourself. The procedure is described in details in the document *AVR053: Calibration of the internal RC oscillator* provided by Atmel.

Serial port and microcontrollers

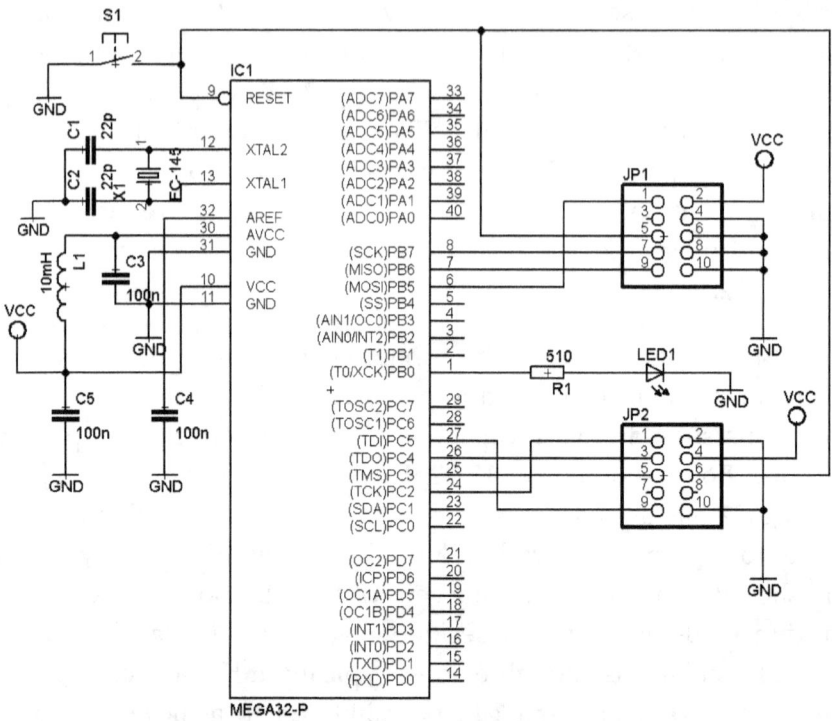

Fig. 3.3.2. Basic ATmega32 circuit.

Connecting a crystal resonator to a microcontroller doesn't automatically make it using it. ATmega32 can have several clock sources: crystal resonator, internal RC oscillator, external RC oscillator and external clock (square wave generator). Selection is made with CKSEL fuse bits. For crystal oscillator faster than 1 MHz the CKSEL bits from 3 to 1 should be set to one of these combinations: 101, 110, 111. Bit 0 should be set to 1. Additionally CKOPT should be set to 0 (programmed). If you want to use internal RC oscillator, set CKSEL bits to 0001 for 1 MHz, 0010 for 2 MHz, 0011 for 4 MHz or 0100 for 8 MHz. By default ATmega32 has internal RC oscillator selected, with frequency set to 1 MHz. When setting fuse bits, refer to documentation of your programming software.

Remember that bit set to 0 is called programmed and set to 1 is called unprogrammed.

Beginners often fail to set fuse bits correctly and it happens that in result they can't access their chips. Very common situation is that they configure a microcontroller to work with an external clock. In this case one has to provide such clock and connect it to pin XTAL1, XTAL2 should be left unconnected. Any simple square wave oscillator will be OK, for example based on Schmitt inverters. Its frequency should be about 1 MHz. Such generator is show on figure 3.3.3.

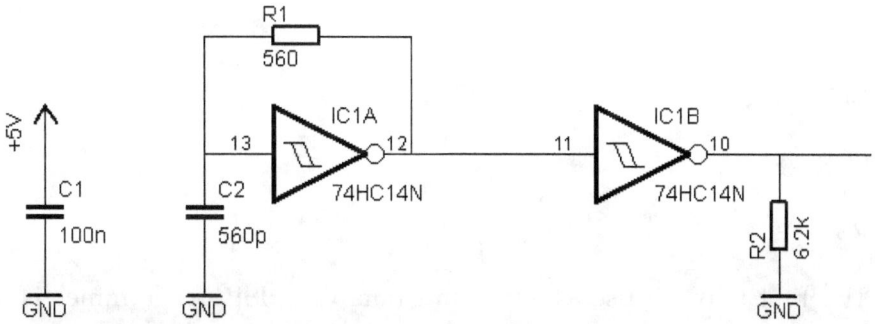

Fig. 3.3.3. Square wave oscillator built with Schmitt inverters.

OK, now we know how to make a microcontroller running, how to reset it and how to provide a clock source for it. What has left is that we need to program a microcontroller somehow. It can be done in two ways: using ISP or JTAG interface. Programming with ISP is much more popular than with JTAG. First of all ISP interfaces are much simpler and less expensive thus they are popular among hobbyists. JTAG interfaces are more complicated and expensive and last but not least only more advanced AVRs, like ATmega32, support JTAG. The advantage of JTAG is that it not only allows programming a microcontroller but allows debugging too. With JTAG you can run your program step by step and observe state of registers and variables.

Serial port and microcontrollers

Let's focus on ISP programming. The matter of required devices was presented in one of former chapters. If you have your ISP programmer purchased or made by yourself it's now time to connect it. According to construction of a programmer one of its ends has to be connected to a PC, usually to parallel or USB port. What about the other end? There are six wires which have to be connected to your microcontroller. First of all we need to connect grounds of the programmer and of the circuit. The programmer is often powered from the same voltage as the circuit so we connect VCC too. There are four wires left. You need to look into a datasheet of you microcontroller to find appropriate pins. For ATmega32 they are:

MISO – 7

MOSI – 6

SCK – 8

RST – 9

The first two are used to transmit data. The ISP programmer is a master device so pin 7 of ATMega32 is used to transmit data to the programmer (Master Input, Slave Output). On pin 6 data go the other way, from programmer to the microcontroller (Master Output, Slave Input). The programmer supplies clocking signal (SCK) to pin 8. Programming is done when a microcontroller is in a reset state so the programmer must provide low logic level to pin 9 during programming. In ATMega32 the reset signal is negated which explains why the microcontroller is in reset state during logic zero.

Having programmer ready we can send our program from PC to a microcontroller, or more precisely, to its flash memory. But wait, we don't have any program written yet. OK, so let's move to software part of our task.

We had many opportunities when it came to choose a development board or a programmer. The same applies to software development tools. Each one of them has some advantages and disadvantages so it's good to install a few of them and test for some time. Maybe it will turn out that it's good to choose two of them and use for different tasks, according to requirements of particular project.

Figure 3.3.2. shows basic circuit for ATmega32 microcontroller. If you want to use crystal, remember to have to it connected, and accompanying capacitors as well. If internal RC oscillator is enough for you, you can omit these elements. Remember that you need crystal if the microcontroller is configured with fuse bits to use it. In such case you won't be able to program the chip without crystal. C3, C4, and C5 are typical 100nF capacitors for filtering supply voltage. In most cases the circuit will work without them but they are recommended as they help to reduce noise. L1 is a choke (coil) which reduces power supply noise for analog-to-digital converter. If you don't use ADC or don't need good accuracy you can remove L1 and connect AVCC directly do VCC. S1 is a reset button. JP1 is a standard ISP connector for programming the chip, and JP2 is a connector for JTAG programming/debugging. Usually only one of these interfaces is used, depending what kind of programmer, ISP or JTAG, is available. R1 and LED1 are an example of a peripheral circuit. Here they are connected to pin 0 of port B but you can use any other free pin of ports A, B, C, and D. You can also connect many kinds of different chips, devices and circuits. For example PC1 (SDA) and PC0 (SCL) pins are for I²C communication. PD1 (TXD) and PD0 (RXD) are used for serial communication and how to use them is the main topic of this book. There are also other special pins: for ADC, SPI, timers and external interrupts which you can use. Some of them we are going to use in examples presented in these books, the other you have to explore by yourself.

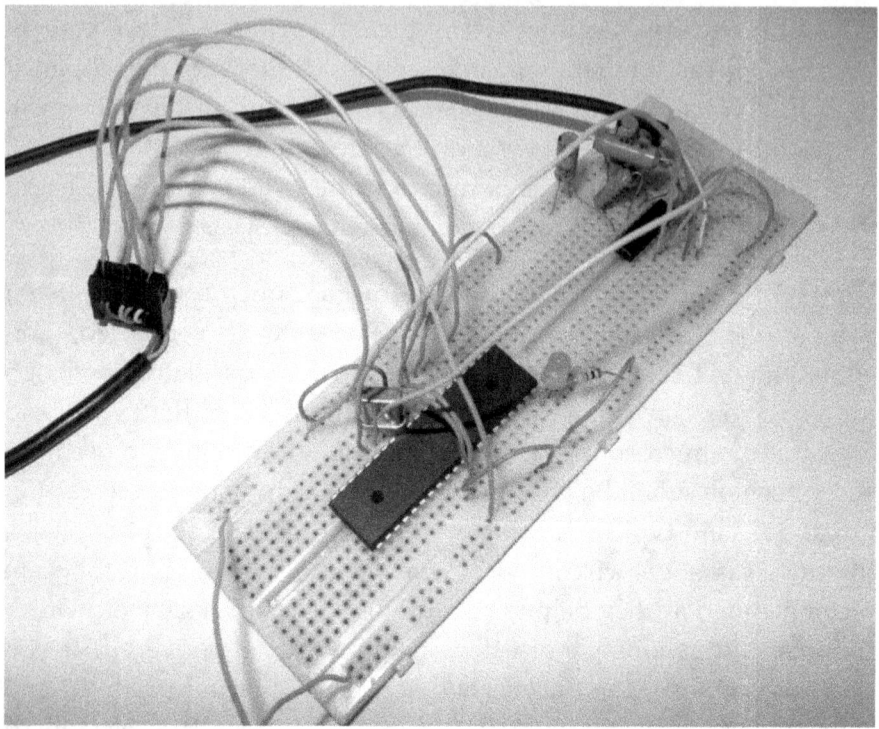

Fig. 3.3.4. Circuit with ATmega32 and MAX232 assembled on a breadboard.

A breadboard is a good choice for experiments with AVR microcontrollers. Fig. 3.3.4 shows simple circuit assembled using such board. There is the ATmega32 microcontroller, crystal with capacitors, LED with a resistor and the MAX232 voltage converter with its resistors for RS-232 communication. Two wires go to 5V power supply, six go to ISP programmer and a 3-wire RS-232 cable goes to a PC.

3.4.   Creating software

3.4.1. Writing program code

Let's write short program. The presented example is very simple but shows a few things specific for microcontrollers programming. When we have it working we will know that we are familiar with the whole process of development and we can easily add more sophisticated functions to our device. Almost all textbooks and Internet coursers about programming start with "Hello world!" programs, simple applications having only one purpose: displaying short text on a screen. In the microcontrollers' world the situation is a bit different. We rarely have any screen and even if we have one, displaying anything on it is definitely not the simplest thing we can do. So what is the most common way of starting an adventure with microcontrollers? Blinking LED. We need to toggle only one bit and the hardware is very minimalistic. That's why a LED was mentioned on a list of elements we need.

So how our simple program can look like? Let's take a look.

```
#define F_CPU 16000000UL

#include <avr/io.h>
#include <util/delay.h>

int main (void) {
      DDRB = _BV(PINB0); //sets pin 0 of port B as output
      while(1) {
            _delay_ms(200);
            PORTB |= _BV(PINB0);
            _delay_ms(200);
            PORTB &= ~_BV(PINB0);
      }
      return 0;
}
```

The program starts with a definition for a compiler. *F_CPU* is defined as 16 million. Why? Speed of the microcontroller is defined by hardware: internal oscillator, crystal resonator, or external generator. This means that if we use different crystal or set another frequency of internal oscillator then the microcontroller will run faster or slower. We have to take it into account when we deal with time measurement. We want our LED to blink which means we need some delays for specific amount of time. As you may guess from the code above, the delay is set to 200 milliseconds. To be sure the value would be really 200 ms we have to provide information about clock rate of the microcontroller. And this is done by definition from first line of the code. If you put there a greater value then the LED will be blinking slower, if you use smaller value then the LED will be blinking faster. U stands for unsigned, L stands for long.

Second line tells the compiler to include file responsible for I/O operations. It allows us to use *DDRx* names to access ports of the microcontroller. Here we use *DDRB* because LED is connected to a pin of port B. The next line is also an include directive. It allows us to use *_delay_ms()* function which holds execution of the program for given amount of milliseconds.

As in programs for Mac or PC, the core routine is called *main()*. It doesn't take any parameters like *argc* or *argv* because there is no operating system which could provide such parameters to the function. Now you can ask: but why does the *main()* function return a value? That's good question. There is no more code besides of the presented one so there is nothing which could receive this *int* value. The answer is simple: we declare *main()* as a function returning value of type *int* because the C programming language standard requires us to do so. Some compilers require the *main()* function to return *int*, some don't. If your compiler allows to have *main()* function not returning anything you can go that way. It doesn't affect execution of a program in a microcontroller in any way.

Now let's analyze the first line of actually executed code. Some value is written to *DDRB*. But what is *DDRB* and what is the value? *DDR* means Data Direction Register. It is used to specify which pins of a port are inputs and which are outputs. There is such a register for each port, *DDRB* is for port *B*. By setting bits in *DDRB* we set operation mode of individual pins of port *B*. Setting a bit to 1 makes a pin an output, 0 makes a pin an input. For example writing 00011111b to *DDRB* makes 3 pins (from 5 to 7) of port *B* to work as inputs and 5 (from 0 to 4) to work as outputs. In any case we need a value which contains appropriate set of ones and zeros. That value can be expressed in many ways. One of them is hexadecimal numbers. If you are familiar with them, you can skip a few following paragraphs.

Suppose you want to make pins 0, 3 and 5 to work as outputs. Then you need to write value 00101001b. In decimal notation it is:

$2^5 + 2^3 + 2^0 = 32 + 8 + 1 = 41$

When programming microcontrollers we often use hexadecimal values. The conversion between binary and hexadecimal is usually easier than between decimal and hexadecimal. When we have binary value we just have to cut the byte in half and calculate values:

$0010b = 2^1 = 2$

$1001b = 2^3 + 2^0 = 8 + 1 = 9$

Thus 00101001b = 41d = 29h. If you want to convert decimal form into hexadecimal you have to think how many 16s there are in a given value, in other words you have to make integer division. Integer part of the result of dividing 41 by 16 is 2. What is left?

$2 \times 16 = 32$

$41 - 32 = 9$

So again we get value 29h. It was mentioned that 255d = FFh but some readers may be not very familiar with hexadecimal numbers so let's explain what F is. Hexadecimal notation uses 16 as a base. The problem is that we don't want to introduce special digits for this notation but it turns out that decimal digits are not enough. There are just 10 of them and we need 16. To solve this problem for numbers from 0 to 9 standard digits are used. For numbers from 10 to 15 letters are used. This way A means 10, B means 11 and so on. Because the base of hexadecimal notation is 16 then the value of maximal digit is 15. As you can check by yourself, 15 is noted as F. Thus 0Fh = 15d, 10h = 16d, and 20h = 32d.

OK, we know how to use hexadecimal numbers. They are not very convenient though. If we want to make pin 5 to work as output we need to write value 32 (20h) to appropriate *DDR* register. This leads us to following line:

```
DDRB = 32;
```

It will work but every time we look at such a line we have to think where 32 comes from. We would like something easier to read. We can use a nice trick:

```
DDRB = (1<<5);
```

What happens here? There is a bit shift operation which moves value 1 (00000001b) by 5 places which gives 00100000b. OK, this looks better, we can see number 5 and *DDRB* on the left so we can easily guess that pin 5 of port B will be working as output. Can this notation be even more friendly? Yes. The libraries give us constants not only for registers or ports but also individual pins. So here is a nicer notation:

```
DDRB = (1<<PINB5);
```

But that's not all. We can get rid of bit shift operator and use following form:

```
DDRB = _BV(PINB5);
```

It's actually the same, we just use a macro for bit shifting instead of using bit shift operator explicitly. The difference is that you don't have to remember how the bit shifting operator works and can use a macro having a name related to bit vectors.

As you can see, there are many ways for writing an expression for particular number value. But what if want to write a binary number, using ones and zeros anyway? In standard C language it's not possible. You can use decimal, octal or hexadecimal numbers, but not binary. Fortunately AVR-GCC contains extensions which make such notation possible:

```
DDRB = 0b00100000;
```

Now you know how the line for configuring *PORTB* works. Let's go further and take a look at some more tricks used often for low level programming. Here we have a line initializing one of registers responsible for serial transmission:

```
UCSRC = _BV(URSEL)|_BV(UCSZ1)|_BV(USBS);
```

This line sets following parameters for RS232: no parity, 7 bits of data, and 2 stop bits. The register which stores the configuration is called *UCSRC* (USART Control and Status Register C). According to the datasheet of ATmega32 we have to set bits of UCSRC in the following manner in order to obtain mentioned configuration:

- for disabled parity checking both *UPM1* and *UPM1* bits should be set to 0

- for 7-bit data transmission *UCSZ2* and *UCSZ0* should be set to 0 but *UCSZ1* should be set to 1

- for 2 stop bits the *USBS* bit should be set to 1

- additionally we need to set the *URSEL* bit because it needs to be set every time *UCSRC* is written. This comes from the fact that the *UCSRC* register shares the same I/O location as the *UBRRH* register and *URSEL* (USART Register Select) bit tells which register is going to be accessed. You have to read the datasheet carefully and pay attentions to such details.

So in the result the *UCSRC* register should have following binary value: 10001100b. To obtain it we perform bitwise OR operation on three bit vectors:

_BV(URSEL) gives 10000000b;

_BV(UCSZ1) gives 00000100b;

_BV(USBS) gives 00001000b

The outcome of bitwise OR operation is 10001100b which is what we wanted. Thus the notation:

```
UCSRC = _BV(URSEL)|_BV(UCSZ1)|_BV(USBS);
```

Now you can guess what happens in the next line of our sample code. Let's recall it:

```
PORTB |= _BV(PINB0);
```

It is equivalent to:

```
PORTB = PORTB | _BV(PINB0);
```

So as a new value of *PORTB* we set the result of OR operation performed on old value of *PORTB* and the bit vector of pin 0. This way we preserve values of other pins of this port, we change just value of one pin (bit). In our example we don't use other pins however and the expression can be simplified:

```
PORTB = _BV(PINB0);
```

This will set bit 5 to 1 and all other bits to 0. But in more sophisticated programs you will be working with more than one pin and when changing one of them you will want to not change the rest of them. Thus it's good to have good habits and remember about other bits.

Further in the code you can see analog line, containing bitwise AND operation and negation. This line puts 0 on pin 0. First of all we perform a negation of all bits returned by the bit vector macro. This gives us the value 11111110b. There is performed a bitwise AND operation with current value of *PORTB*. This way all bits of port B keep their old values except of bit 0 which is set to 0.

The remaining lines which we haven't explained yet are calls to a library function which introduces a delay, given in milliseconds. Pin operations and delays are in a loop so everything is constantly repeated and connected LED can blink constantly, until power is down. Actually almost all programs for microcontrollers contain main loop which is executed infinitely.

Important note: there are different ways of reading and writing states of pins of a microcontroller. In our example, when we were setting the value of *PORTB*, we were reading it in order to preserve values of remaining bits. But we were not reading states of pins of the microcontroller! Don't read *PORTn* to check the logic level on some pins. *PORTn* are registers storing states of pins configured as outputs. If you read *PORTn* you will get the value you previously wrote to it, no matter what are the voltages on pins of particular port. The input values are stored in *PINn* registers, e.g. *PINB*.

In presented short example we don't use interrupts but a few words about them must be said. This is not a book strictly about AVR programming and we are not going to dig into details but there is one important thing which must be brought up. Most popular AVRs: ATtiny and ATmega are 8-bit microcontrollers. This means that they operate on single bytes, even if there is a 16- or 32-bit variable

Serial port and microcontrollers

declared in our program. If there is such a variable, eg. uint16_t, then it is processed in more than one step. We say that operations on that variable are not atomic. Should we be concerned about this fact? Usually not, but if we use interrupts the situation may become more complicated. The problem is that an interrupt may happen in the middle of a non-atomic operation. If the operation and the interrupt handler are referring to the same variable then your program may work in a way you didn't expect. How interrupts work, what are their sources, etc. you can find in many books and on many websites and we are not going to cover this topic here. But the issue of non-atomic operations is rarely presented although bugs related to this problem may lead to hard and time consuming debugging.

If you encounter a problem caused by an interrupt triggered during non-atomic operation you can try several things. One of them is to rewrite the code in such a way that a 1-byte variable is enough instead of 2-byte or 4-byte. The other solution is to change the algorithm so the troublesome variable is eliminated or the interrupt is not used. Or we can use a critical section. It's a piece of code which is guaranteed to be executed without interruptions (atomically). The avr-gcc compiler provides the *ATOMIC_BLOCK(type)* macro for this purpose. How does it work? It simply turns interrupts off so execution of the embodied code can't be interrupted. The *type* parameter can be one of two values: *ATOMIC_RESTORESTATE* or *ATOMIC_FORCEON*. In the first case state of interrupts is restored to the original one, in the second interrupts are turned on. Let's take a look at piece of code which illustrates the problem and solution:

```
counter = 300;
while(counter!=0);
```

The *counter* variable is a global volatile 16-bit variable constantly decremented by timer interrupt:

```
ISR(TIMER0_OVF_vect)
{
  counter--;
}
```

As you can guess this way we can introduce a delay in our program. For example if the interrupt is fired every 1 millisecond the delay is 0.3 s. What is the problem? Consider the moment when the value of *counter* is 256 (0100h) and the condition of *while* loop is checked. The program checks the less significant byte which is 0 (0000h). And before it can check the more significant part of the variable the interrupt occurs. In interrupt's handler the variable is decremented to 255 (00FFh). When the handler's execution is finished program returns to checking the variable. Lower byte is already checked so it checks the high byte. Now it is 0, exactly as the low byte was when it was checked. And if both less significant and more significant parts of the variable were checked to be zero then it is concluded that the variable as a whole is also zero and the *while* loop breaks. This way they delay is shorter than it was intended to be. Worse, such a situation happens randomly, depending when the interrupt occurs. As it was said there can be several solutions. For example in this case we can use a 1-byte variable and make two 150 ms delays, one after another. Or we can configure the timer to run two times slower if it doesn't affect other parts of the program. If we want to use a critical section we can do it this way:

```
counter = 300;
volatile uint16_t tmp;
do {
     ATOMIC_BLOCK(ATOMIC_FORCEON) {
        tmp = counter;
     }
} while (tmp != 0);
```

As you can see the solution is to use a second variable to be checked by the *while* loop. It is copied inside a critical section so it is not affected by non-atomic access. The additional variable should be declared as volatile so it is not optimized by a compiler. From the optimization point of view the *tmp* variable is redundant as it stores the same value as *counter*. Of course, as you see, in practice this is not always true. But the compiler doesn't "know" that an interrupt can modify the variable and can perform optimization resulting in broken code.

The *ATOMIC_BLOCK* macro works by disabling interrupts for the time of execution of instruction located within it. But you don't need to be afraid that some interrupts will be lost, they will be fired as soon as the interrupt flag is set again. Just keep the code inside critical section and in interrupt handler as short as possible so second interrupt of the same kind won't happen before processing of the first one is finished.

3.4.2. Compilation and programming

We have the code analyzed so it's time to compile it. If you use Atmel Studio you can just press F7. If you like to work with WinAVR's command line you can run *avr-gcc* command. The effect of the compilation is a set of files. One of them has extension .elf which tells us that it's an Executable and Linkable Format file. It's an executable file format popular on Unix-like operating systems, including Linux. It's also used for AVRs. It contains executable code but also other data, like symbol tables and other debug information. However we are not interested in this file now. What we need is a file we could supply for a programmer. This file has extension .hex and is in Intel HEX format. It contains actual executable code with additional data suitable for transmission, such as addresses and checksums. All data are written in hexadecimal format so can be transferred over 7-bit links. They are also more human readable than in a binary form.

HEX files are supported by many applications for programming microcontrollers. One such program is ISP Programmer described in one of previous chapters. When you run it, click Setup button and select STK200/300 Evaluation Board. In the main window set the speed to 1 000 000 Hz. If everything is properly connected then click Read signature to check if programmer can communicate with the chip. Checking the signature is required to enable buttons for reading and programming a microcontroller. If a microcontroller has been correctly recognized you can select HEX file with your program. Before you click Pgm Flash button, you have to erase flash memory. So click Erase and then Pgm Flash. When programming is done you can click Verify Flash to check if programming was done correctly. To speed things up you can use Erase & Program All button. This button also turns of reset so the program is started when programming is finished. If everything is OK you should see blinking diode. Restart button can be used to reset a microcontroller by turning reset on for a short time.

If you use Atmel Studio then programming is very easy. When you run your program, by pressing F5 or using *Debug* menu, the microcontroller is automatically programmed. Of course a programming interface supported by Atmel Studio must be connected. You can also use *Tools -> Device Programming* menu to program chip's Flash memory or fuse bits.

Serial port and microcontrollers

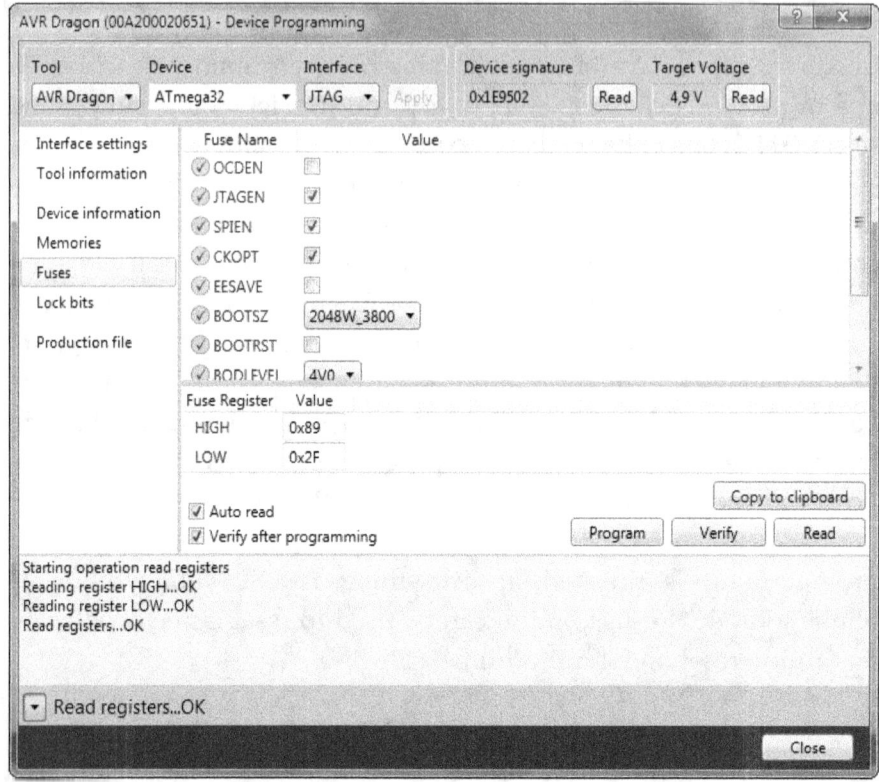

Fig. 3.4.2.1. Atmel Studio programming tool

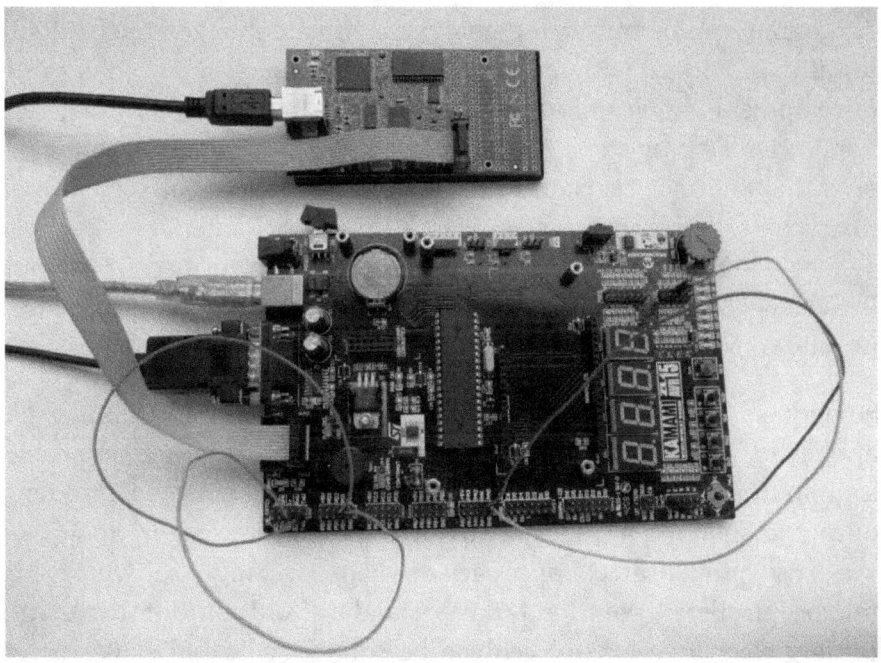

Fig. 3.4.2.2. ZL15AVR development board and AVR Dragon programmer working as JTAG interface

3.4.3. Debugging and testing

If your microcontroller doesn't show any signs of life despite of Flash memory being correctly programmed, you probably made some trivial mistake when connecting external circuits. Check voltages on pins of the microcontroller and verify correctness of connections.

The situation is usually more complicated if you see that your microcontroller is working but not the way you wanted. Probably there is some bug in your code, you have to find it and remove. This can be done in two ways, similarly as when debugging applications for PCs or Macs.

Serial port and microcontrollers

A program can be traced and analyzed with a debugger. Such a tool, usually provided with a compiler or included in an integrated development environment, gives a possibility of running an application step by step. You can stop a program and check contents of variables and registers, you can observe the call stack too. Microcontrollers can be debugged using JTAG interface which is available in more advanced AVRs, for example in ATmega32 but not in ATmega8. Of course it requires appropriate hardware interface, usually connected to a computer with USB.

Sometimes programmers use other ways to trace execution of an application. If a debugger is not available they use available input/output functionality such as printing something on standard output, displaying a message box or writing a message in a log file. We use the same approach when writing software for microcontrollers. We don't have standard output but there are usually some free pins which can be connected to LEDs. We don't have a computer screen but sometimes there is some LCD display available, enough for debug messages. We don't have files (OK, we can use SD cards but it's not very convenient for log files) but we can use serial port to send any data to a PC. As you can see lack of JTAG interface doesn't leave you without possibilities of debugging.

Some people say that a programmer who knows debugging tools very well is not a good programmer. Good programmer knows how his/her program works and doesn't need to use a debugger. This is partially true. It's good to think for a while before starting typing a source code. Consider different ways of performing the same task. Sometimes there are a few algorithms to choose and one is better than the other. Sometimes you come up with a simple solution which turns out to be working in 99% of cases. Then you struggle for many hours with this one percent when the program fails. On the other hand it's almost impossible to write flawless code, working perfectly after first run. Datasheets for microcontrollers are quite big and it's impossible to remember every detail of their internals working. They also require a little different approach than when writing programs for PCs or Macs. Anyway it's always good to think the problem over, read datasheet the microcontroller and documentation of used functions, and try to find situations where troubles potentially can happen.

There are a few things you have to consider when programming microcontrollers. For example there may be in interrupt caused by external event which stops execution of main code and causes interrupt routine to be executed. After this routine is finished, main program continues. Make sure that it won't trigger any unwanted behavior, e.g. interrupt handler code causing unintended modification of areas of memory used by main code.

The other problem you may encounter is if you don't check what a function you use actually does. Let's say there is a function for sending a byte through serial port. You write a line which calls this function to send one byte. Then you write second such line to send another byte. When you run the program you notice that only one byte is received by the other device while it should receive two bytes. Why? It turns out this function in fact can be used to send data but transmitting a byte is not what it actually does. So how does it work? All it does it putting the byte in a buffer. After this simple operation the microcontroller executes next line of code. After the byte is written to the buffer an USART circuit integrated inside the microcontroller starts to send it. Hence transmission is done in parallel with execution of following instructions. This is why two calls of a function for sending data cause only one byte to be transmitted. The second call overwrites the value written by first call before it is sent by USART. Solution for this issue is simple, before each next call there must a piece of code waiting for end of transmission initiated by previous call.

Atmel Studio makes debugging quite easy. You can start and pause (break) execution of a program using appropriate toolbar buttons,

Serial port and microcontrollers

Debug menu and keyboard. Very useful function is a step-by-step execution. You can execute a line of code using F10 key. Current line, the one which is to be executed, is marked with a yellow arrow. You can also use breakpoints. To add or remove a breakpoint click desired line of code and press F9. If you run the program the execution will stop at the line with the breakpoint. When execution is stopped you can inspect microcontroller's memory, registers, call stack, and variables.

```
int main() {
    uint8_t stateOld = 3;
    init_UART();
    sei();
    DDRA |= _BV(DDA0);
    PORTA |= _BV(      );
    for(;;) {
        state = PINA & _BV(PINA1);
        if (state!=stateOld) {
            stateOld = state;
            while(!(UCSRA & _BV(UDRE)));
            if(state==0) UDR = 'P'; else UDR = 'N';
            _delay_ms(50);
        }
    }
}
```

Fig. 3.4.3.1. Current line and a breakpoint during execution of program under control of Atmel Studio environment

114

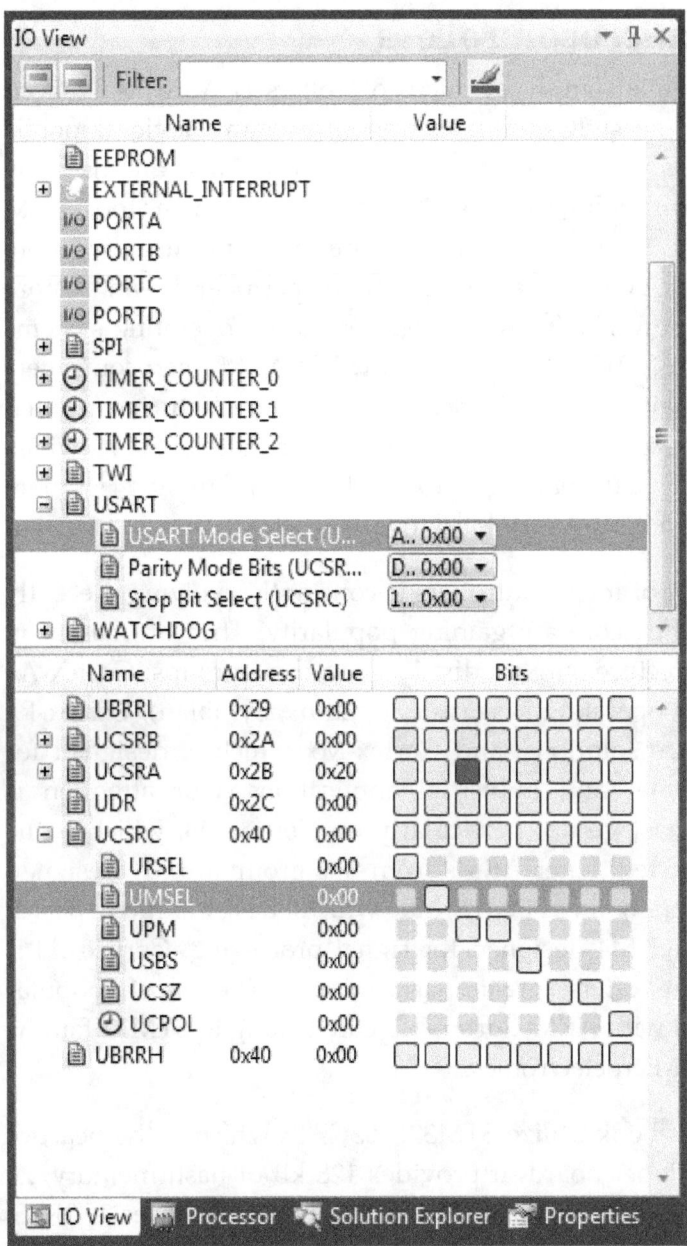

Fig. 3.4.3.2. I/O registers view during debugging session in Atmel Studio

4. STM32 microcontrollers

The other presented family of microcontrollers is ARM (Advanced RISC Machine). These ICs are now present in many devices: mobile phones, wireless routers, embedded solutions. They are used by professionals but hobbyists are starting to use them too. ARM microcontrollers have 32-bit architecture and are usually more powerful than AVRs. They offer more peripherals and more computational power. Often they are used to run operating systems like Linux or FreeRTOS. The price is low but ARM microcontrollers are provided mainly in SMD packages and manufacturing printed circuit board (PCBs) for them is more expensive than for AVRs which are provided in DIP packages too. Also programming is more difficult although it's not very hard to learn.

The ARM family of microcontrollers is constantly evolving. Recently ARMs with Cortex core are gaining popularity. They are based on ARMv7M architecture and are divided into three groups: Cortex-Ax for devices with operating systems (e.g. Linux, Symbian), Cortex-Rx for time sensitive solutions, and Cortex-Mx which is designed for low cost consumer and industrial applications. Our attention is focused on the last group, particularly on Cortex-M3, which is the first one constructed. Later ARMs from this group are M1 designed for implementation in FPGA, M0 which has minimal power consumption, and M4 designed for signal processing. Cortex-M3 is inexpensive but offers great performance. The most popular microcontrollers with ARM Cortex-M3 core belong to STM32 family, produced by STMicroelectronics.

Examples in this book utilize STM32F103CBT6 which is the heart of ZL30ARM developer board. It provides 128 kB of flash memory, 20 kB of SRAM, two SPI interfaces, two I²C interfaces, three UARTS, USB interface, CAN interface and ADC (analog to digital converter).

4.1. Development tools for STM32 microcontrollers

4.1.1. STM32 Standard Peripheral Library

For writing software for Cortex-M3 ARM microcontrollers one doesn't actually need any additional libraries. However development without such help is not very convenient. Hard to remember registry names, poor portability and necessity of writing own routines for accessing peripherals shows that some standard library would be much desired.

The STMicroelectronics company provides such a standard library for its STM32 microcontrollers. It is called STM32 Standard Peripheral Library and is compatible with CMSIS (Cortex Microcontroller Software Interface Standard). The library provides easy access to peripherals included in STM32 ICs. Thanks to CMSIS compliance a programmer gets a vendor-independent hardware abstraction layer. It has two important features: makes porting software from one STM32 microcontroller to other one a quite easy task and secondly allows focusing more on higher level functionality than on internal structure and working of a processor. The library comes with many examples showing how to use it in practice.

On the other side there are also disadvantages of the STM32 Standard Peripheral Library. This piece of software is not perfect and sometimes its code is not optimal or it's even buggy. It causes some overhead too, which makes your code bigger and slower than it could be. Anyway its learning curve is not so big and the library does its job. As you gather more knowledge about STM32 microcontrollers you can adapt the library to your needs or even abandon it. You are encouraged very much to examine how the functions the library provides work. It gives a good understanding how STM32 programming looks like and serves as a starting point to write own routines, more optimized and more flexible than provided by the library. Examples presented in this book make use of the STM Standard Peripheral Library.

The STM32 Standard Peripheral Library is available for download on STMicroelectronics website. It is constantly updated so it's good to have the newest version. The library is not hard to use but you must remember about a few things.

Source files are located in two subfolders of *Libraries* folder: *CMSIS* and *STM32F10x_StdPeriph_Driver*. You must configure your compiler to look for C header files in appropriate subdirectories:

- CMSIS\CM3\DeviceSupport\ST\STM32F10x
- CMSIS\CM3\CoreSupport
- STM32F10x_StdPeriph_Driver\inc

The first one contains *stm32f10x.h* header file and the second one contains *core_cm3.h*. The third one is a folder with functions for various peripherals. Remark: in IAR Embedded Workbench for ARM 6.20 or newer don't include the *CMSIS\CM3\CoreSupport* directory by go to project options and in *General Options* select *Library Configuration* tab and select *Use CMSIS* checkbox. That's because the compiler that comes with EWARM 6.20.1 and later has got support for more intrinsics than previous versions, which leads to an incompatibility issue with CMSIS source code that is included in projects.

The other important thing is the stm32f10x_conf.h file. It determines what peripherals header files are included (e.g. *stm32f10x_gpio.h*) and also contains definition of *assert_param()* macro which many library files use. To have stm32f10x_conf.h included in your project, the *USE_STDPERIPH_DRIVER* symbol must be defined. To have it defined look at compiler configuration. In IAR Embedded Workbench open project properties and go to *C/C++ Compiler* section. Click the *Preprocessor* tab and type *USE_STDPERIPH_DRIVER* in

Defined symbols field. The stm32f10x_conf.h header file is included just like other files which means that the compiler needs to know how to find it. Thus you have to add the location of this file to list of paths searched when looking for included files. If you forget about it then the compiler may just fail or include another file of the same name. For example Keil μVision is provided with STM32 header files, including stm32f10x_conf.h. If you don't configure the μVision compiler to include your file it will include that file. OK, but where is the *stm32f10x_conf.h* file located? You can find it in *Project\STM32F10x_StdPeriph_Template* subdirectory of the STM32 Standard Peripheral Library. If you are going to modify this file, copy it to you project's directory, this way other your projects, which include the file from the *STM32F10x_StdPeriph_Template* directory won't be affected. To configure IAR Embedded Workbench to search for that header file in project directory add *$PROJ_DIR$* to the list of include directories.

If you wish you can create your project without the *stm32f10x_conf.h* file. If you look at source codes available on the Internet you can notice that such approach is not uncommon. Of course this means that what was provided by the *stm32f10x_conf.h* file must find its place in other files of your project. You have to define the *assert_param()* macro and remember to include header files of the peripherals you are going to use in you project.

Some functions from the library work differently depending of the SMT32 family you use, e.g. *medium density*, *high density value*, etc. Therefore you need to define in your project an appropriate symbol, e.g. *STM32F10X_MD* for *medium density* microcontroller. You can find the list of available symbols at the beginning of *stm32f10x.h* file.

The other important thing is that the microcontroller has to know how to handle interrupts. We need to configure the interrupt vector table to point to appropriate functions. This is done with so called startup file, usually written in assembler. Startup files for popular

development environments and for each of the STM32 lines are located in
Libraries\CMSIS\CM3\DeviceSupport\ST\STM32F10x\startup
directory. Find the one suitable for your environment and chip and add to your project.

Bear in mind, that the startup file provided by STM32 defines the reset handler in the way that the first thing performed after reset is execution of *SystemInit()* function. This function is defined in *system_stm32f10x.c* file located in *Libraries\CMSIS\CM3\DeviceSupport\ST\STM32F10x* directory. In general it is responsible for setting up the system clock, by default to 72 MHz. Again you have two choices: you can add that file to your project or write your own *SystemInit()* function.

The entry point of the library is stm32f10x.h header file so it should be included in your source code. You should also not forget about implementation files of the library (*.c). You have header files (*.h) included but the implementation files should be compiled too so remember about adding them to your project. When all required .c files are compiled then linker can link them into one executable.

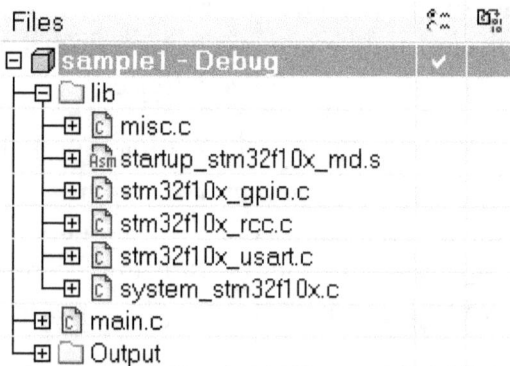

Fig. 4.1.1.1. Example project structure in IAR Embedded Workbench. lib is a group in the project, not a directory. It's a place for library files.

4.1.2. CrossWorks for ARM

CrossWorks for ARM is a development suite from Rowley Associates, available for Windows, Linux, Mac OS X, and Solaris. It consists of integrated development environment called CrossStudio, C, C++ and Assembler toolchain from GCC, CrossWorks C Library (non-GPL and non-LGPL), and CrossWorks Tasking Library. CrossWorks for ARM supports number of JTAG interfaces, including simple and popular Wiggler and many devices based on FT2232 chip. Rowley Associates offers own JTAG interface, called CrossConnect for ARM. It is of course supported too and is most compatible with CrossWorks for ARM.

CrossWorks for ARM is commercial software and its price depends on licensing. The lowest price is for personal, non-commercial use and the highest is for commercial use. In any case the feature set is the same. If you want to try the software, you can download 30-day evaluation version.

4.1.3. Keil MDK-ARM Microcontroller Development Kit

The Keil company provides MDK-ARM which is a complete software development environment for Cortex™-M, Cortex-R4, ARM7™ and ARM9™ processor-based devices. Microcontroller Development Kit contains μVision IDE, C/C++ Compilation Toolchain, Keil RTX – deterministic, small footprint real-time operating system (with source code), USB Device and USB Host stacks are provided with standard driver classes, Complete GUI Library for embedded systems with graphical user interfaces, and other useful tools and libraries. MDK-ARM is available in four editions: MDK-Lite, MDK-Basic, MDK-Standard, and MDK-Professional. All editions provide a complete C/C++ development environment and MDK-Professional includes extensive middleware libraries. MDK-Lite (32KB) Edition is available for download. It does not require a serial number or license key.

4.1.4. IAR Embedded Workbench for ARM

An interesting commercial IDE is IAR Embedded Workbench for ARM. It supports wide range of ARM cores: ARM7, ARM9, ARM9E, ARM10E, ARM11, SecurCore, Cortex M0, M1, M3, M4, R4(F), A5 and XScale and comes with over 2400 example projects. It can work with JTAG interfaces such as ST-LINK, Stellaris FTDI or Macraigor OCDemon. It works also with IAR's own devices: J-Trace, J-Link, and J-Link Ultra. All of the three support Cortex-M3 chips. IAR Embedded Workbench for ARM includes own C/C++ compiler and linker. It has also good support for many real-time operating systems. There are two free version: one is time limited (30 days) and the other is code size limited (32 kB). The second one is called KickStart and is especially interesting as 32 kB is enough for many projects. Example projects for this book were created using this version.

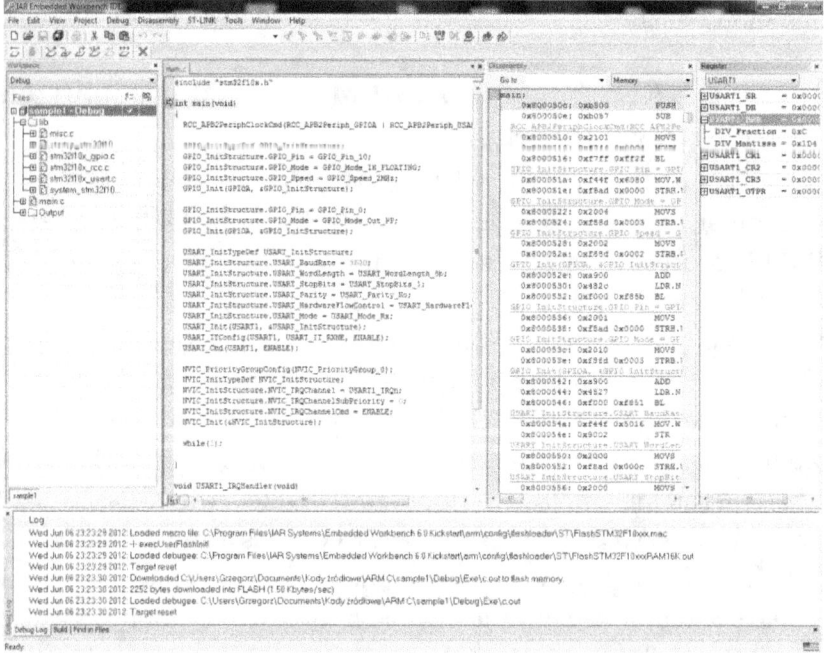

Fig. 4.1.4.1. IAR Embedded Workbench

4.1.5. Raisonance Ride7

Ride is an acronym for Raisonance Integrated Development Environment. As for time of writing this book it was available in version 7. You can download it from Raisonance website but you must know that it's just IDE. For any development possible you need to have a so called RKit which contains a compiler, libraries, and other required tools. There are several RKits available, the one for ARM microcontrollers it called RKit-ARM. Like WinARM, it makes use of AVR GCC, which is included and you don't have to install it separately. Both Ride and RKit are available for free. The disadvantage is that the only supported debugger/programmer is RLink, also provided by Raisonance. It is offered in two versions: Standard and Professional. The first one is cheaper but supports programs up to 32 Kbytes which is usually enough. Of course you can use Ride just for writing source code and compiling it. Chip programming can be done with third party software and hardware.

4.1.6. Atollic TrueSTUDIO

The Atollic TrueSTUDIO development environment comes in two versions: Professional and Lite. As you may guess, the Professional version has more features but requires the purchase of license. The Lite version is available for free but has lower functionality. The most important limitation of the free version is support for just one JTAG interface: ST-LINK. Although the Lite version can be used without any cost, registration is required to install the program. Moreover user is often nagged about upgrading to paid version. Nonetheless TrueSTUDIO Lite can be a good choice. Like the commercial version, it's based on popular Eclipse environment and GCC compiler. Built-in wizard helps to easily create a new project, only little configuration is needed.

4.1.7. WinARM

WinARM is an Open Source toolkit created by Martin Thomas. It's similar to WinAVR, both are packages of free programs and are based on GNU GCC compiler. The difference is that AVRs are produced by Atmel and there are many similarities between AVR microcontrollers. On the other hand ARMs are manufactured by various companies and their variety is much higher. This makes ARM tools more complicated. Moreover ARM microcontrollers need startup code which initializes them after reset. Such code is written in assembly language and linked to main program, which is usually written in C. But overall using WinARM is analogous to using WinAVR. There are command line tools, like arm-gcc.exe compiler and make.exe tool which processes Makefile scripts. They are accompanied by many additional programs, including Programmers Notepad editor, H-JTAG or GNU GDB.

If you started looking for WinARM to download, you may be surprised by the fact that the newest stable version is 20060606 and newest test versions are 20070505 and 20080331. At least that's how the situation looked like in the end of 2010. If you encounter some problems caused by outdated WinARM components, you can try to upgrade to some of test versions. Another solution is to take GNU tools from other packages, like Codesourcery's Sourcery G++ Lite Edition for ARM, devkitARM, or Yagarto.

4.1.8. OpenOCD

One of the most popular debuggers is GDB (GNU Debugger). As part of GNU project it is free and open. It is used by many development tools, including WinARM. It can be used as well for debugging programs executed on ARM microcontrollers. But it can't work if it is run on a PC and the program runs on a chip. In this case it needs GDB server, an application which can connect to target microcontroller. OpenOCD (Open On-Chip Debugger) is such software. It uses JTAG interface to connect to a microcontroller and serves as an agent between a chip and GDB running on a computer. Then debugging can be done with any development environment which supports GDB, for example Eclipse. OpenOCD can work with

JTAG interfaces based on FT2232, RLINK, JLINK, Wiggler and many more. It is available on Windows, BSD, and Linux.

Usually OpenOCD receives commands from GDB but we can connect to it using some telnet client and issue commands directly. For example we can type *reset* to reset our chip or type *poll* to see state of the core of the chip. The other interesting command is *reg* which dumps content of registers. To stop execution of a program we can type *halt*. The command *step* causes execution of one instruction from the program.

4.2.   Hardware tools for programming STM32 microcontrollers

4.2.1. Wiggler

JTAG interfaces are usually connected to a computer through USB. The Wiggler interface from Macraigor Systems utilizes a parallel port. Its design is very simple which makes Wiggler very easy to build and makes the cost low. There are many clones of Wiggler so there is no problem with buying it if you don't intend to assemble it by yourself. The Wiggler clone provided by Kamami is known as ZL14PRG. It works with HiTOP-Link, OCD Commander, IAR Embedded Workbench for ARM, and CrossStudio from Rowley. Wiggler is also used with Open Source tools like WinARM and OpenOCD. While Wiggler is simple and inexpensive it has also its limitations which makes it nowadays not very good choice. First of all it's slow which makes programming and debugging problematic. The other important issue is usage of parallel port which not available on many computers now. It's mentioned here for historical context. Anyway parallel port JTAG interfaces are still used for some microcontrollers.

4.2.2. ST-LINK and ST-LINK/V2

ST-LINK is a low-cost in-circuit debugger/programmer for the STM8 and STM32 microcontrollers. It connects to application or evaluation boards for programming and debugging via STMicroelectronics' SWIM single-wire connection for the STM8 and via a JTAG standard connection for STM32 chips. It communicates with a PC through USB. It is supported by popular programming environments: Keil uVision, IAR Embedded Workbench, Atollic TrueSTUDIO, STMicroelectronics STVP, and STMicroelectronics STM32 ST-Link Utility.

Fig. 4.2.2.1. ZL30PRG JTAG interface

There is also a newer version of the tool called ST-LINK/V2. ST-LINK is offered by the Kamami company as ZL30PRG and ST-LINK/V2 as ZL30PRGv2. For both devices there is software called ST-LINK Utility, provided by STMicroelectronics. It has such features as Flash memory reading and programming, setting option bytes, displaying chip information, and upgrading programmer's firmware.

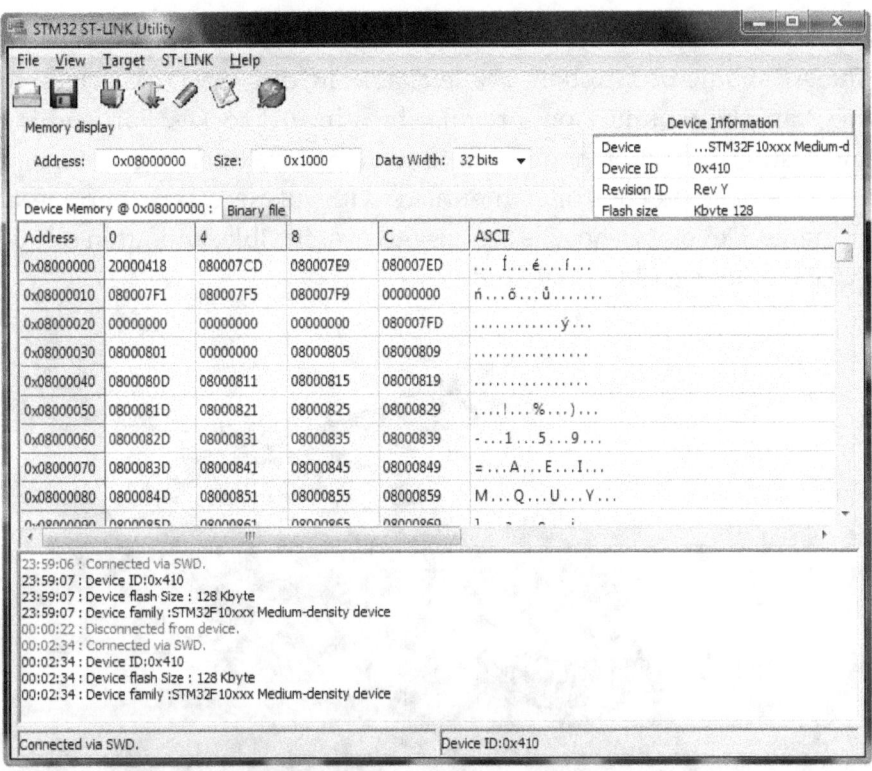

Fig. 4.2.2.2. STM32 ST-LINK Utility

4.2.3. FT2232-based JTAGs

Quite large group of JTAG interfaces are devices based on FT2232 chip from FTDI. This chip is USB <-> dual UART converter. It's fast and has many features, such as bit-banging or Multi-Protocol Synchronous Serial Engine. It contains also built-in EEPROM memory which can be used to program the chip. As it is USB-USART converter, it is seen by Windows operating system as virtual COM port. JTAG interfaces which are equipped with FT2232 are supported by OpenOCD, CrossWorks, and some other development environments. They are relatively inexpensive and easy to build. Some of FT2232-based JTAGs are: Turtelizer 2, OOCDLink, Amontec JTAGkey2, and Olimex ARM-USB-OCD.

4.2.4. ZL30ARM developer board

The ARM microcontrollers are available in tiny, SMD packages so they can't be mounted on a regular breadboard socket. For learning and testing purposes one needs to have a developer board. Such a PCB has a microcontroller mounted with all necessary peripheral elements. Developer boards are easy to use, flexible, and often can be expanded using additional modules.

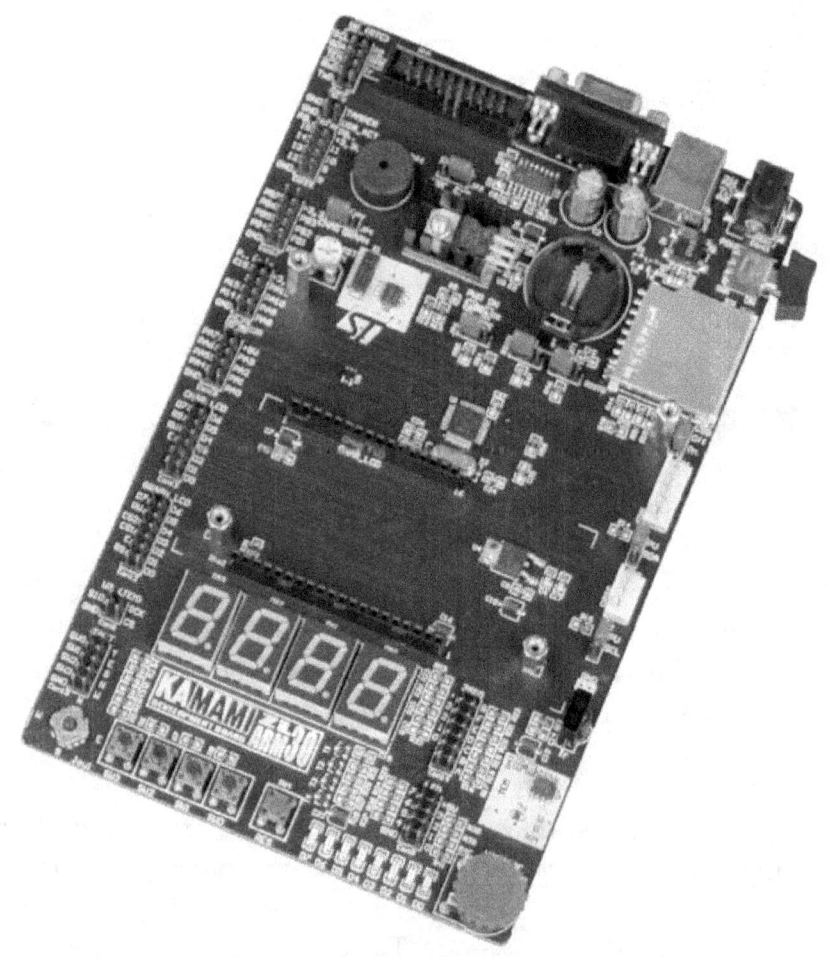

Fig. 4.2.4.1. ZL30ARM development board

The Kamami Internet store offers many different developer boards. One of them is ZL30ARM. It is equipped with STM32F103CBT6 ARM Cortex-M3 microcontroller which has 128 kB of Flash and 20 kB of SRAM. It also has many interesting peripherals, including 2xSPI, 2xI2C, 3xUART, USB, CAN, and ADC. The board has 8 LEDs, 4 buttons, a joystick, TC77 digital thermometer, STLM analog thermometer, TSOP31236 IRED receiver, a potentiometer, a slot for SD cards, a USB connector, M41T00 real time clock, a piezo buzzer, 4-digit LED display and connectors for LCD displays (graphical and alphanumeric).

4.3. *Designing own hardware for STM32s*

STM32s are available in small SMD packages so it's not possible to easily attach them to a breadboard socket. Thus software development and experiments are usually done using developer boards, e.g. ZL30ARM. If you are designing a PCB for your own device you have to connect the microcontroller correctly. Let's consider the STM32F103CBT6 chip which is used in ZL30ARM. If you use other ARM Cortex-M3 microcontroller, refer to its datasheet for information about pinout.

The supply voltage range for STM32F103CBT6 is from 2 to 3.6 volts. The chip is typically powered from 3.3V power source. The often used voltage regulator providing such voltage is LM1117. There are several power pins you have to connect. VSS_1, VSS_2, and VSS_3 (pins 23, 35, and 47) have to be connected to ground. VDD_1, VDD_2, and VDD_3 (pins 24, 36, and 48) have to be connected to positive supply voltage. There are two pins delivering power for ADC, reset blocks, RCs and PLL, these are VSSA (pin 8) and VDDA (pin 9). They have to be connected to power lines as well, VSSA to ground and VDDA to supply voltage. The microcontroller allows sustaining the real time clock, 32 kHz clock and backup registers using a battery when the main power is not present. Thank to this the chip can resume normal operations when main power comes back. If you want to use this function you have to connect the positive contact to the VBAT pin (it's pin 1), negative to the ground. The battery voltage must be in range from 1.8 to 3.6V.

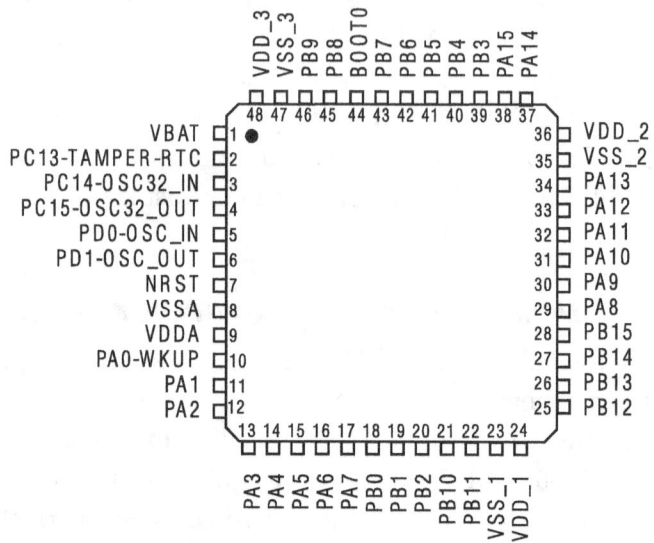

Fig. 4.3.1. Pinout of STM32F103CBT6 in LQFP package.

As for the clocking signal, it can come from several sources, similarly to AVRs. The situation is a little more complicated though. Generally there are four clock signal sources: HSI (High Speed Internal), HSE (High Speed External), LSI (Low Speed Internal), and LSE (Low Speed External). HSI utilizes internal 8 MHz RC oscillator. Its accuracy is ±1% at 25°C but can be as bad as ±3% in whole temperature range. On the other hand HSE is as accurate as external signal source. There can be two kinds of sources: a crystal oscillator or a generator providing sine, square or triangle signal. The crystal can have the frequency from 4 MHz to 16 MHz, usually it's 8 MHz. In case of external signal it can be from 0 to 25 MHz, typically 8 MHz. The crystal oscillator is a popular solution. If you want to use it, connect it to pins OSC_IN and OSC_OUT (pins 5 and 6). Like in case of AVRs, it has to be accompanied by small ceramic capacitors connected between each of the two pins and ground. Their value must be in range 5 – 25 pF, typically 22 pF. When the microcontroller starts, it uses HSI clock. If you want to use HSE, you have to programmatically turn it on. The chip can work with a frequency higher than the one of the clocking source. There is a PLL (Phase-Locked Loop) which can multiply the base clock (either internal or external) to the maximum system clock frequency of 72MHz. You must remember that the higher the clock the faster the chip but also the higher power consumption.

HSI and HSE can be used as a system clock. HSE, LSE, and LSI can work as real time clock and LSI can work as real time clock or watchdog clock. For LSE working as RTC usually the 32768 Hz crystal oscillator is used. It has to be connected to pins OSC32_IN and OSC32_OUT (pins 3 and 4). Accompanying capacitors should be in range from 5 pF to 15 pF. LSI uses internal RC oscillator, about 40 kHz. It is very inaccurate however; its actual frequency can be from 30 to 60 kHz. Fortunately the RTC prescaler can be calibrated for better accuracy. The procedure is described by ST in the application note AN2604. The advantage of LSI is very low power consumption.

Similarly to AVRs, Cortex-M3 microcontrollers have integrated power-on reset (POR)/power-down reset (PDR) circuitry. It ensures that the chip will start correctly after power on and that it will be reset when supply voltage drops below certain level. This threshold can be configured using *PLS[2:0]* bits of the *PWR_CR* register. Reset is triggered by low logical level so if you want to perform it manually, add a button between NRST pin (it's pin number 7) and ground. It is recommended to add a 100 nF capacitor between NRST and ground to protect against parasitic resets.

The next two pins you have to take care of in your projects are BOOT0 and BOOT1 (pin 44 and 20). They select the memory from which the microcontroller boots. If the pin BOOT0 is connected to the ground (logical zero) then after reset the microcontroller starts to execute a program stored in its flash memory. In this situation the state of BOOT1 doesn't matter. If BOOT0 is connected to VCC (logical one) then there are two possibilities. If BOOT1 is 0 then the microcontroller executes a bootloader program stored in its ROM memory. If BOOT1 is 1 then the SRAM memory is used. The most popular scenarios are booting from Flash or starting the bootloader. In practice selection of one of them is made depending whether there is a JTAG interface available. If you have JTAG you use it for both debugging and programming. With JTAG sending the compiled code to a chip is very easy and fast. If JTAG interface is not available, you can run the bootloader and send your program using serial cable connected to USART1 pins of your chip. As you can see bootloader offers you less expensive way of programming – you don't have to buy a JTAG interface which definitely costs more than simple RS232 cable. On the other hand it's less convenient because programming takes more time and you have to reconfigure boot pins each time. Reconfiguration is required because when you ran the bootloader and used it for writing a program into Flash memory; you want to execute your program from that memory. That means that during debugging and testing of your program you end up constantly switching between two boot modes. bootloader and Flash. Hence as a developer you will probably use JTAG and have booting set always to Flash. If you create devices for other people, who are less familiar with electronics but would like to get upgrades of the software you created for the chip used in the device, you would like to let them use bootloader for firmware upgrades. Thus a good idea is to use jumpers on a PCB for easy switching between boot options. Detailed description of STM32 bootloader can be found in application note AN2606.

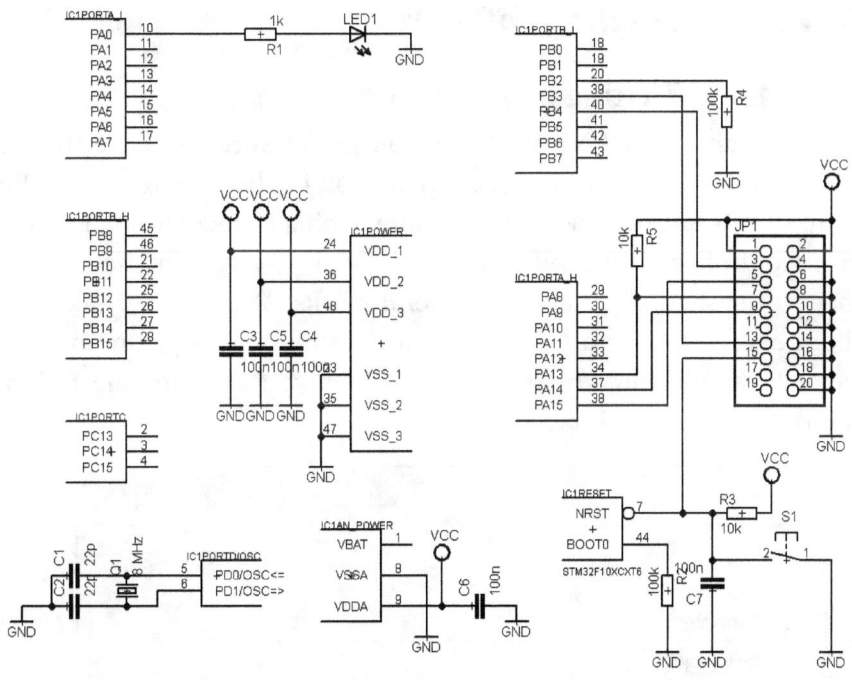

Fig. 4.3.2. Basic STM32 circuit. JP1 is a standard 20-pin JTAG connector.

4.4. Creating software

4.4.1. Writing program code

Programming for ARMs isn't hard but is for sure less easy than in case of AVR microcontrollers, especially on the beginning. One of the more significant reasons is much more sophisticated core of an ARM microcontroller, requiring more initialization, including setting the speed of system clock and peripheral clocks. You have to take care about a vector table and linker configuration, although many IDEs have it usually already preconfigured or you can use the STM32 Standard Peripheral Library for that.

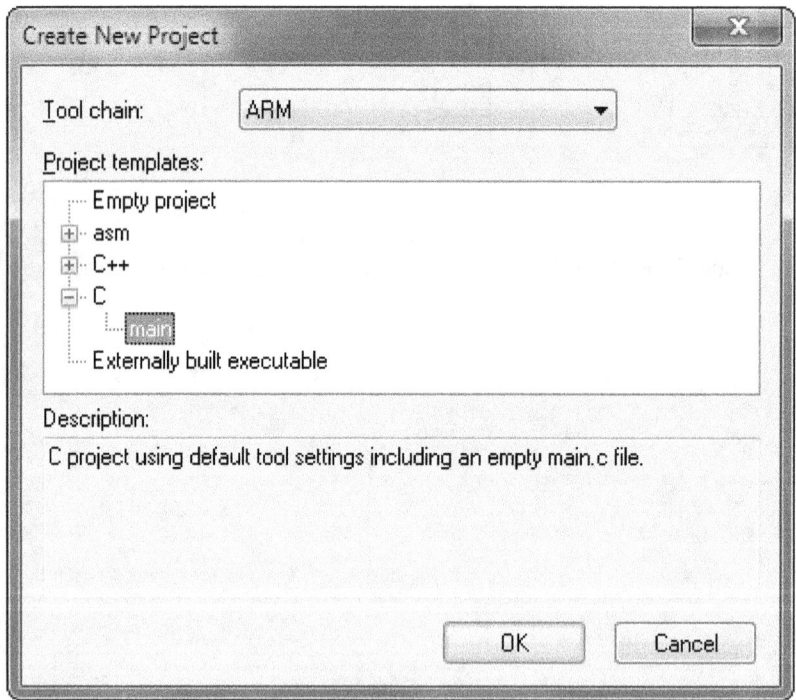

Fig. 4.4.1.1. Creating a new project in IAR Embedded Workbench

In examples presented in this book we take advantage of the STM32 Standard Peripheral Library so we need to have it properly configured in our project. This includes defining appropriate symbols in compiler options, setting library's paths in compiler configuration and adding startup file with interrupt vectors table. This is described in details in the chapter devoted to the library. Selecting a target for a compiler looks a little bit different in different IDEs, in IAR you need to select "ST STM32F10xxB" (fig. 4.4.1.2). Screenshots from 4.4.1.3 to 4.4.1.x. show settings related to the STM32 Standard Peripheral Library.

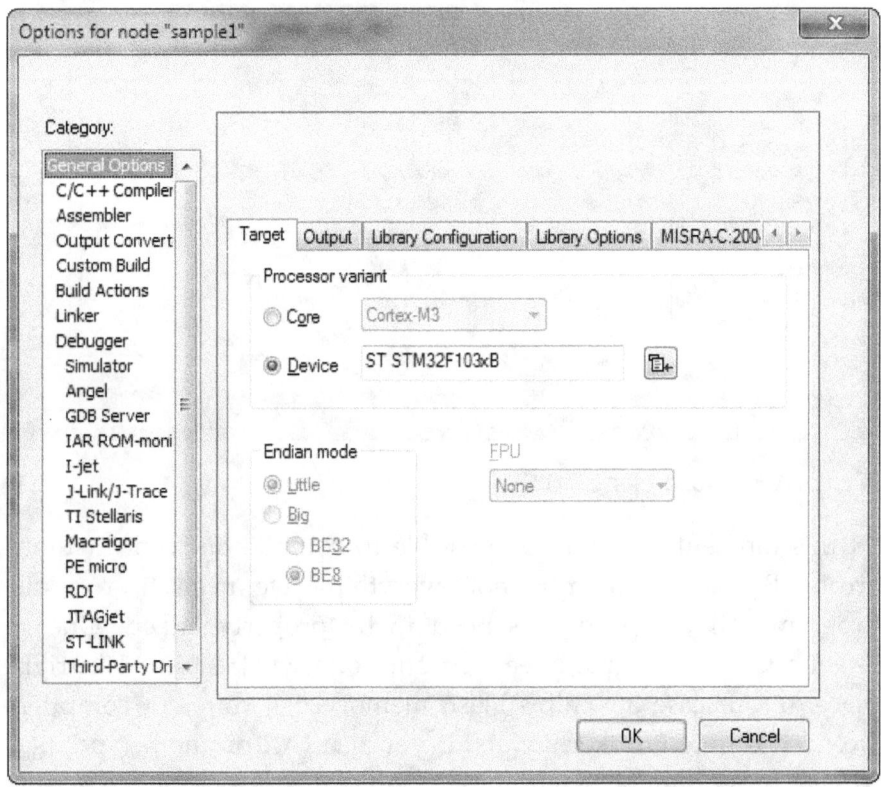

Fig. 4.4.1.2. Microcontroller type selection

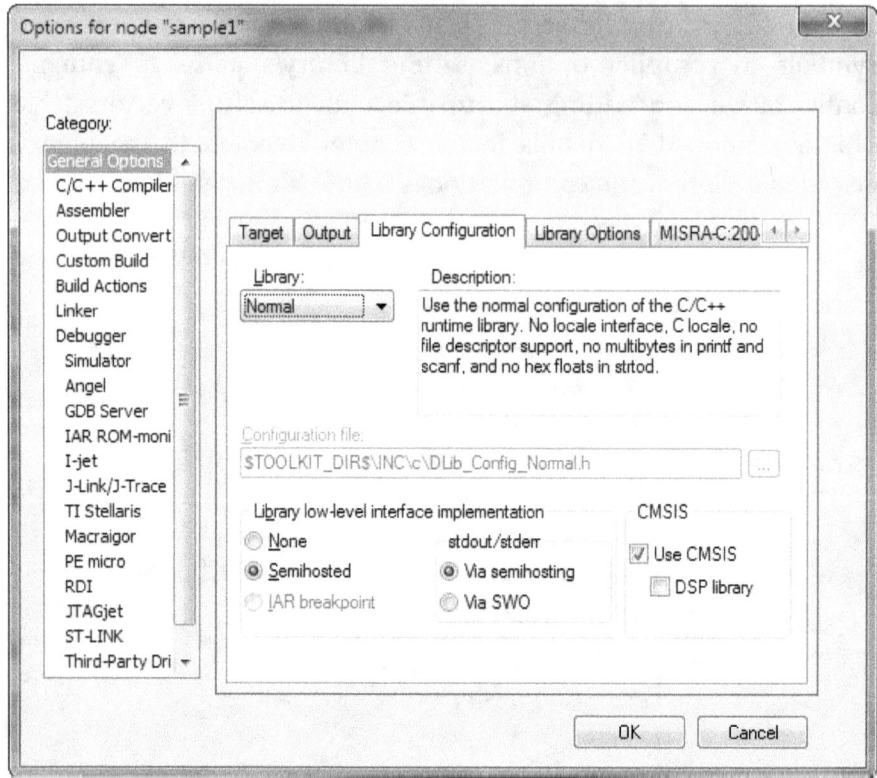

Fig. 4.4.1.3. Support for CMSIS

At this moment it should be possible to compile our code without errors. But compilation is not everything we need to run our program. All compiled files need to be properly linked. One of linker's tasks is to guarantee that our code will be placed in right place. In our case it's chip's Flash memory. For different compilers we need different linker scripts. If you use IAR or another popular STM32 programming environment, you can use one of the scripts provided the STM32 Peripheral Library. You can find them in the templates subdirectory. For example the linker script for IAR Embedded Workbench for AMR is located in *Project\STM32F10x_StdPeriph_Template\EWARM*. Linker scripts are usually provided with development tools, you can find them on the Internet as well.

As for the startup code, which for example defines interrupt handlers, the situation is similar. Startup files for IAR are located in *Libraries\CMSIS\CM3\DeviceSupport\ST\STM32F10x\startup\iar*.

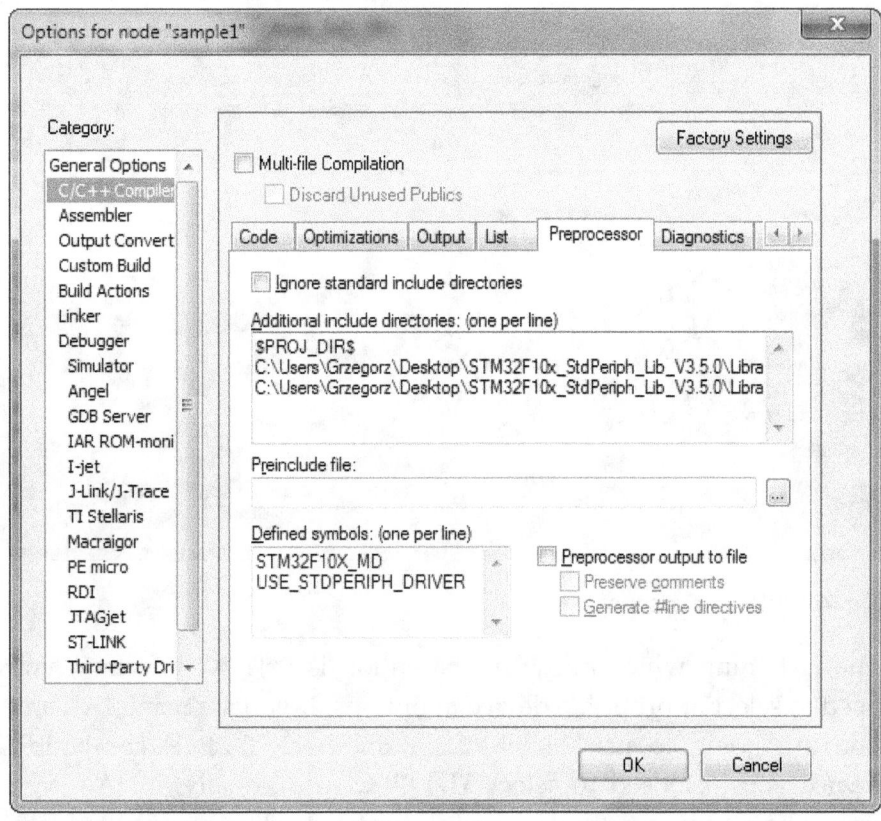

Fig. 4.4.1.4. Preprocessor options

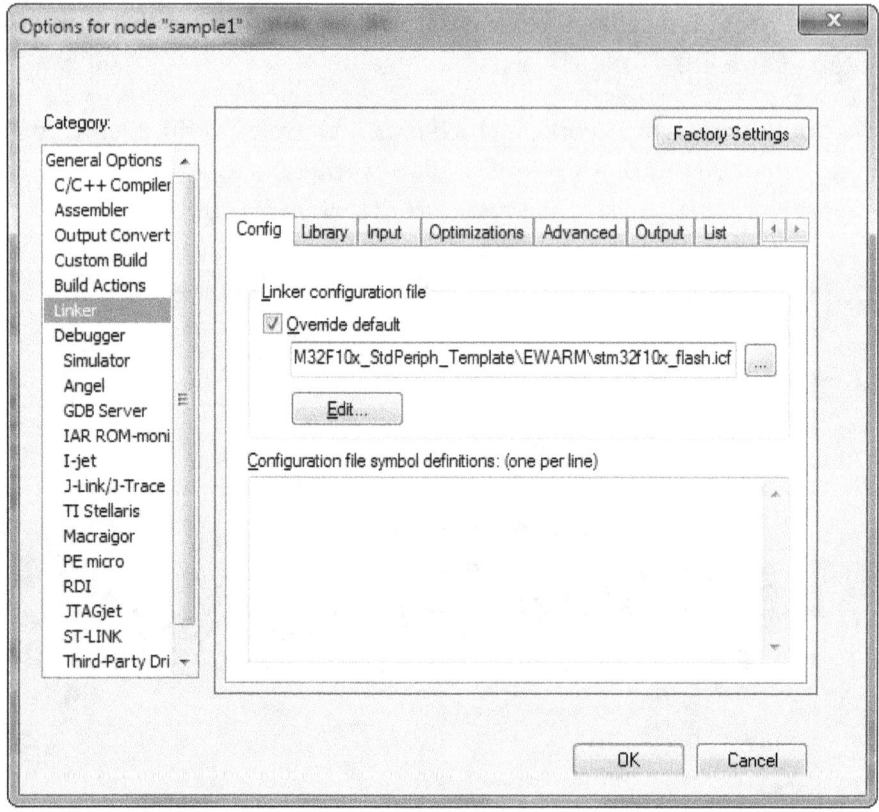

Fig. 4.4.1.5. Linker script

The last thing which needs our attention is a JTAG interface, you need to select appropriate device in options. If you are using IAR and you are going to use ZL30PRG or another ST-LINK compatible device you just need to select ST-LINK. Additionally, in IAR, you need to be sure that the *Use flash loader(s)* checkbox is selected. We don't want to use our JTAG just for debugging but for programming the chip's Flash memory as well. In other IDEs similar configuration may be required.

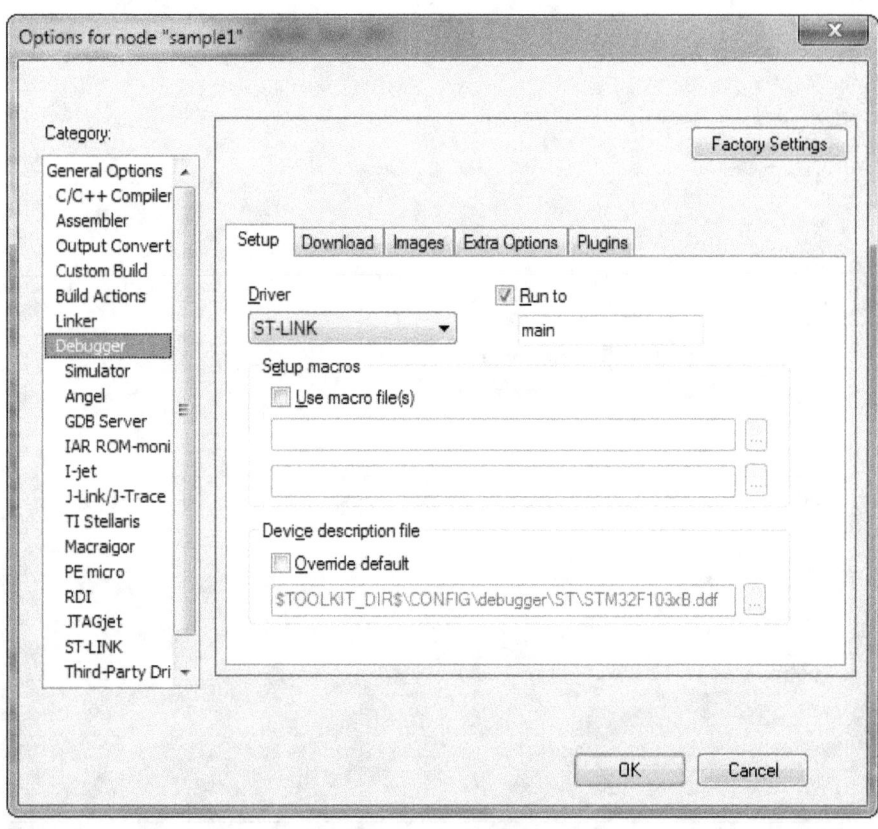

Fig. 4.4.1.6. Programming/debugging tool selection

Serial port and microcontrollers

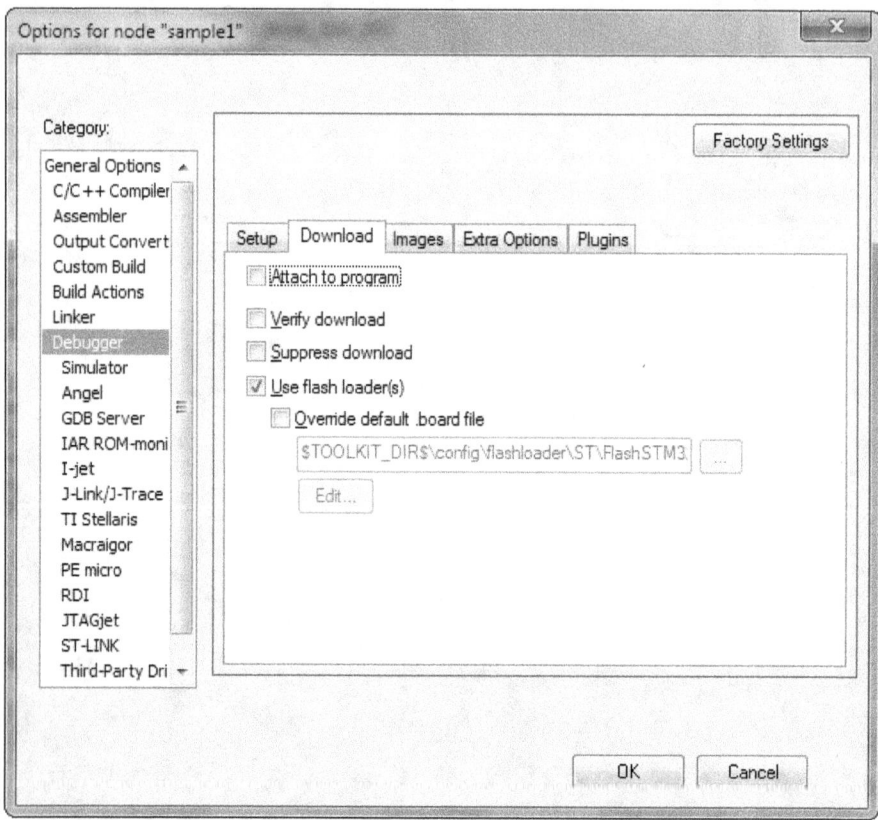

Fig. 4.4.1.7. Flash loader enabled

Finally we can look at sample code:

```
#include "stm32f10x.h"

void Delay(volatile unsigned int delay);

int main()
{
  RCC_APB2PeriphClockCmd(RCC_APB2Periph_GPIOA, ENABLE);

  GPIO_InitTypeDef GPIO_InitStructure;
  GPIO_InitStructure.GPIO_Pin = GPIO_Pin_0;
  GPIO_InitStructure.GPIO_Mode = GPIO_Mode_Out_PP;
  GPIO_InitStructure.GPIO_Speed = GPIO_Speed_2MHz;
```

```
GPIO_Init(GPIOA, &GPIO_InitStructure);

while(1) {
  GPIO_SetBits(GPIOA, GPIO_Pin_0);
  Delay(2000000);
  GPIO_ResetBits(GPIOA, GPIO_Pin_0);
  Delay(2000000);
}

}

void Delay(volatile unsigned int delay)
{
  for(;delay>0;delay--);
}
```

First of all we include the *stm32f10x.h* file which is an entry point of the STM32 Standard Peripheral Library. We are not including the *stm32f10x_conf.h* file explicitly because it is included by the library. The next thing in our code is declaration of a simple function which realizes a delay. This function is defined at the end of the code, we are going to analyze it later.

The next few lines deal with configuration of an I/O port. This is one of differences between STM32 and AVR. On AVR you just had to configure a pin as an input or output, here the situation is more complicated. Peripheral circuits in STM32 microcontrollers need a clocking signal to operate. Thus in order to use a peripheral circuit, such as I/O port, we have to turn clocking signal for it. Here, using the *RCC_APB2PeriphClockCmd()* function we are enabling clock for port A. You may wonder if there is a *RCC_APB1PeriphClockCmd()* function. Yes, there is. This is a result of the fact, that there are two peripheral buses: low speed Advanced Peripheral Bus (APB1) and high speed Advanced Peripheral Bus (APB2). The first one can work with speed up to 36 MHz and the second one at 72 MHz. General purpose I/O ports are connected to APB2, this is why we use *RCC_APB2PeriphClockCmd()*.

When the clock is enabled we configure the port itself. We do it using a structure of type *GPIO_InitTypeDef*. With the *GPIO_Pin* field we select port's pin to configure, here it is pin 0. Then we set *GPIO_Mode* field to *GPIO_Mode_Out_PP*. As you can guess it configures the pin to work as output in push-pull mode. Push-pull means that an output can work as either source or sink of current. The other possible mode is open-drain (*GPIO_Mode_Out_OD*), where the output can work only as a sink for current. Let's say you want to connect a LED to a pin. With push-pull you can connect your LED between a pin and either VCC or GND. As long as you take care about proper polarity of LED and appropriate resistor to limit current such a setup can work. If the LED is connected between a pin and VCC (LED's cathode to the pin, anode to VCC, with the resistor of course), you can turn it on with logical 0 on the pin. In case of plugging the LED between a pin and GND you can turn it on with logical 1 (LED's cathode to GND, anode to pin, still remember about serial resistor to protect the LED). With open drain only one combination is possible: LED connected between pin and VCC. If you connect LED to GND it won't work as the output pin can't work as current source. The next thing we set in our code is speed of the pin, here it is 2 MHz. Other possible values are 10 and 50 MHz. Probably you would like to use highest possible value but remember that the faster the pin the steeper slopes of generated signals. Steep slopes contribute to high frequency noise which could be an unpleasant effect. Thus you should keep the speed of I/O pins low. When all interesting fields are set we call the *GPIO_Init()* function to perform actual configuration of a port, here it is port A.

Now the configuration part is finished and we can move to blinking a LED. We do it by periodically setting ones and zeros on an I/O pin. To set one we use *GPIO_SetBits()* function and to set zero we use *GPIO_ResetBits()*. As you see the first argument is port and the second one is pin.

The last thing which needs an explanation is the *Delay()* function. It takes a value and decrements it until it reaches zero. The higher the value the longer it takes to reach zero. But what values should we use, why the value used in the example is 2 million? To answer this question we need to know how the *for* loop was compiled. Every good development tool includes disassembler so it's easy to check out. In our example the loop is compiled into three instructions: *SUBS*, *CMP* and *BNE.N*. The first one decrements the value using subtraction. Then using *CMP* the value is compared with zero. *BNE.N* checks the result. If it was not equal to zero then it jumps to the first instruction. *SUBS* and *CMP* take one clock tick each while *BNE.N* need two clock ticks/cycles (an exception is when comparison was positive and *BNE.N* doesn't jump, then it need only one clock cycle). Hence there are 4 clock cycles required for one iteration. As the chip's internal clock is 8 MHz we need 2 million iterations to get a delay of one second.

But there is a catch. There may be 5 o 6 clock cycles needed for one iteration if the microcontroller works at higher frequencies and so called waitstates are used for accessing Flash memory. In such case jumps need more clock cycles and the loop's iteration is executed longer. Moreover some compilers, depending on optimization options, can remove the loop completely as in fact it doesn't perform any operations. Therefore it's better to use timer based delays. However, if you need to quickly write short program then loop based delay is OK.

This was very simple program, meant to show just basics of programming for ARM microcontrollers. Our goal is not to discuss every aspect of writing programs for those chips but to focus only on serial communication. There are however some things which need to be explained deeper because they are going to be important in majority of your programs. I'm going to say a few words about selecting clock source and multiplying its frequency.

Presented code takes advantage of internal high-speed clock (HSI – High Speed Internal) which is a default clock source, used after chip's reset. As it was said, its speed is 8 MHz. Sometimes however, we need a different speed or want to have a more accurate clock source. This is why we often use crystal oscillators for external clock (HSE – High Speed External). If there is a crystal oscillator connected to your microcontroller and you want to use it, you can do it with following short code:

```
RCC_DeInit();
RCC_HSEConfig(RCC_HSE_ON);
RCC_WaitForHSEStartUp();
```

On the beginning we set clock configuration to default. Then we enable HSE and wait for it to be ready.

Another thing we can do is increasing frequency of system clock to be higher than frequency of clock source. This way our microcontroller can work at up to 72 MHz using 8 MHz crystal oscillator. Higher speeds are required for such functions as USB or Ethernet communication. You must remember however that higher frequency of operation causes higher power consumption. Thus we keep the frequency as low as possible.

There is also another effect occurring at higher frequencies. The internal Flash memory where your code is stored is not very fast, it can work up to 24 MHz. At higher frequencies it requires waitstates. Without them instructions wouldn't be correctly fetched from the memory. As you can guess slow speed of Flash slows down execution of the program because instructions can't be fetched in every processor's cycle. Fortunately there is a prefetch buffer which significantly reduces effects of slow Flash. It works very well but has a problem with branches (jumps) as it has to predict where the execution flow is going to go. That's the cause of the problem with delay loops which was mentioned above; the length of delay they introduce is not directly related to system clock speed.

4.4.2. Compilation, programming and debugging

All the operations mentioned in this chapter's title can be usually performed using the same IDE that the code was written with. For example IAR Embedded Workbench comes with own compiler and supports several JTAG interface for programming and debugging. In general the process is very similar to developing for AVR microcontrollers. The code has to be written, compiled, sent to the chip and debugged if the program doesn't behave as expected. The difference is that for STM32 microcontrollers we use usually JTAG interfaces, not ISP programmers, so debugging is easier.

5. Using serial port on a PC

Before we say something about serial communication on a microcontroller it would be good to talk over the PC part. This way when it will come to programming a microcontroller we will have required knowledge and applications.

In many cases our device will require custom application working on a computer. This application will be responsible for receiving and interpreting data from a microcontroller and also sending commands and data to the device. But sometimes and specifically during tests the most suitable tool would be a serial terminal. Users of Windows operating systems from 95 to XP can use built-in application called HyperTerminal.

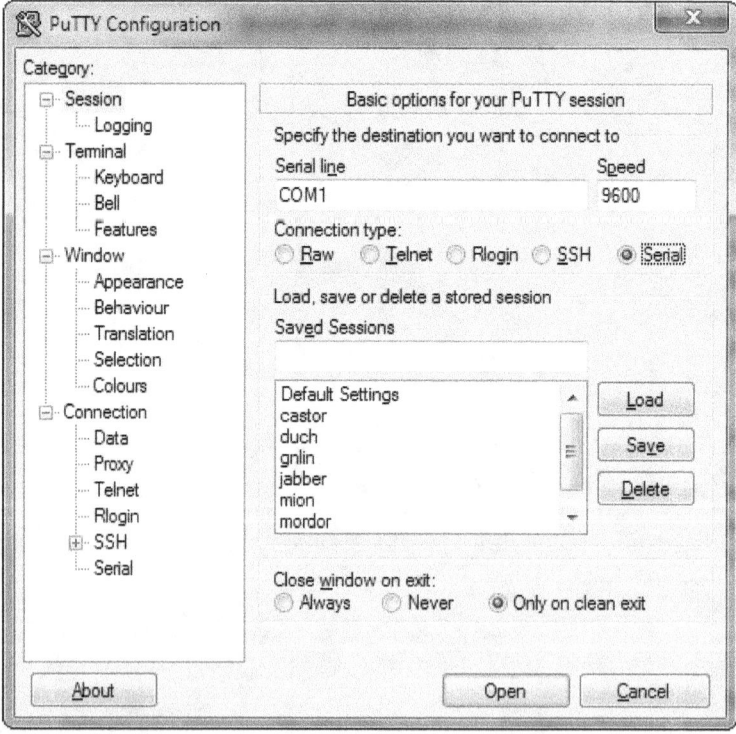

Fig. 5.1. PuTTY configuration window.

The other interesting application is PuTTY. It is most known as SSH client but it can also work as a terminal with serial port. On main screen it is required to select *Serial* option instead of default *SSH*. Now you can specify a serial port you want to use. To set other properties of RS-232 communication navigate in the tree on the left to *Connection* and then to *Serial*. When you make experiments using PuTTY keep in mind that when you hit Enter in PuTTY's terminal window only the CR (carriage return) character is being sent, the LF (line feed) is not.

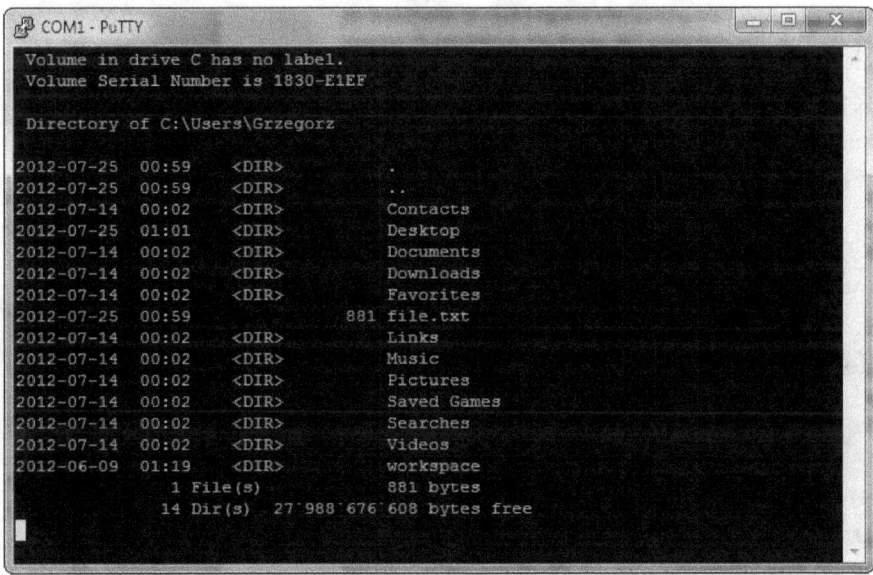

Fig. 5.2. PuTTY terminal window. Visible output from the dir command redirected through RS-232 from another computer.

There are also many other applications which are useful for working with serial ports. In most cases our device will transmit data in some particular format and would also require data encoded in some way. In such case general purpose terminal application won't be suitable and we have to write custom software for communication with our device. It's not hard and almost any programming language would be good for this task.

5.1. Programming serial port in .NET

Let's start with C# on Windows. It's one of the easiest ways of developing an application capable of transferring data over RS-232. You don't even need to install any development environment. It's enough if you have .NET Framework installed, which is true for most Windows machines. The framework includes a C# compiler (*csc.exe*) and you can write the source code in any text editor, even in Notepad. Of course it would be better to have some programming suite, for example Visual Studio. Free Express edition should be enough. If you have Visual Studio installed, launch it and create new C# project. As a template select Windows Forms Application. Type some name for your project and select path to store source code files. Visual Studio creates default project with empty window. This window is called Form1 and will be a base for our program.

Dealing with serial port with .NET is not a hard task. All we need it a System.IO.Ports namespace so let's declare its usage. Open source code of the window and add following line, next to other declarations of used namespaces:

```
using System.IO.Ports;
```

In order to send and transmit data with RS232 we need to instantiate an object of class SerialPort. We can to this inside the class of the window:

```
static SerialPort ComPort = new SerialPort("COM1");
```

As you can see we provide a name of serial port as the parameter for the constructor of SerialPort object. There are several constructors available which allow specifying other number of parameters. Let's stick with this constructor and set other properties in following lines. But wait, we can't do this in the same place where the declaration was placed. We want to have our code executed when the program starts. As we use Form1 as main structure of our program, let's take advantage of constructor of Form1 object. We can add our code below *InitializeComponent();* line. What we have to do is to specify parameters of RS232 transmission:

```
ComPort.BaudRate = 57600;
ComPort.Parity = Parity.None;
ComPort.StopBits = StopBits.One;
ComPort.DataBits = 8;
ComPort.Handshake = Handshake.None;
```

As you can see it's very easy. The baud rate is 57 600 bits per second. There is no parity but there is one stop bit. We use 8 bits of data and no handshake. At this point all the required parameters of RS232 transmission are set. Data transmission can be performed with another RS-232 compatible device but some more settings are needed to make further programming easier.

Now it's possible to call methods responsible for receiving data. The problem lies in fact that they are synchronous. This means that wait for data to come and execution of a thread is stopped. Sometimes it's OK but usually we want to get data asynchronously and to have our program notified about incoming data. In order to handle incoming data there must be a function which will be called each time any bytes are received. Such a handler can look like this:

```
private void OnSerialDataReceived(object sender,
SerialDataReceivedEventArgs args) {

}
```

It's empty now but it may be already specified as the data handler. SerialPort class has a property called DataReceived which is a multicast delegate. It stores pointers to functions which have to be invoked when data are received. We add such functions using += operator:

```
ComPort.DataReceived += OnSerialDataReceived;
```

Before we add some code to the handler we have to think for a while about how it should work. It's obvious that we want to read data from the port's input buffer. But what data and what to do with them? SerialPort class has several methods for accessing received data.

One of them is *Read()* which copies data from input buffer to a given one. Actually there are two such methods because *Read()* method is overloaded and one version is designed to work with array of Byte and the other one with array of Char values. The first parameter is a buffer the data will be copied to. The second parameter is an offset. We are not always interested in writing from the beginning of the buffer. In many cases the buffer will contain previous data and we don't want to overwrite them. Thus the offset will be helpful. The third parameter determines number of bytes to be copied. We use this value when we want to read only some amount of data. The *Read()* method returns number of bytes copied from input buffer. Here is an important thing. If third parameter is larger than amount of data in input buffer then the *Read()* method will wait for them. This means that execution of application (or more precisely the reading thread) will be stopped until more data are received or timeout occurs. Thus usually it will be necessary to know how many bytes were received and read only that amount. We can check it using the *BytesToRead* property.

If you need to read only one byte, there are two methods available: *ReadByte()* and *ReadChar()*. Like in previous case, they work in the same way but differ only in type of returned data. They also wait for data if none were received.

There are two methods designed for handling text data. *ReadExisting()* reads available bytes and returns string value. It uses current encoding to create string from bytes. If the input buffer contains some lead bytes at the end they may remain in the buffer after the method finishes. Lead bytes are special sequences used in some encodings, for example UTF-8. *ReadLine()* also works with text data. It tries to read whole line from input buffer, i.e. it reads bytes until it receives new line sequence. This sequence can be <CR><LF>, <CR> or <LF>. <CR> is an ASCII character meaning carriage return and is a byte of value 13 (0Dh). It is also written as an escape sequence '\r'. Similarly <LF> means line feed and is a byte of value 10 (0Ah) and is written as an escape sequence '\n'. Different operating systems use different combinations of these characters. On Windows it's <CR> and <LF> and on Linux and Mac OS X just <LF>. On some systems, like old versions of Mac OS new line in text is marked with <CR> character. When you write software for interchanging text data you have to know what standard is used by both devices. *ReadLine()* method returns when it encounter any of mentioned sequences and doesn't put them into returned string. You must remember that when <CR> character is received then the method doesn't wait for <LF>. Normally it's not a problem but may cause some troubles when you mix text and binary reads. For example if a line was ended with <CR><LF> and you call *ReadByte()* after *ReadLine()* then you will get <LF> (byte of value 13).

The last of methods for receiving data is *ReadTo()*. It is similar to *ReadLine()* because it also reads input data up to some sequence of characters. But in this case we can exactly specify this sequence. It is passed as a string value. *ReadTo()* reads data up to this sequence and returns them as string. The specified sequence is not places in that string but is removed from the input buffer.

Writing to serial port is easier than reading. We don't need to create any handler and list of methods for writing is shorter. There is an overloaded method *Write()* which can be used to send string or content of a buffer. When string value is transmitted, *Write()* method takes current encoding into account. By default it's ASCIIEncoding which means that characters of codes greater than 127 are not transmitted. Instead the method sends bytes of value 63 ('?'

character). Other available encodings are UTF-8, UTF-32 and Unicode.

Two more versions of *Write()* method can be used to send contents of a buffer. It can be array of byte type values or char type values. The first case is suitable for sending binary data, we can be sure that our data will be send as they are, without any modification. In case of character values the *Write()* method applies encoding to transmitted data using the same principles as during transmission of strings.

There is one more method for sending data, text data to be precise. It's *WriteLine()* which takes string as an argument and operates in similar way as for character values. It differs by the fact that when all data are sent it sends additional string containing new line marker. As you already know, new line may be described using different combination of bytes/characters. *SerialPort* class has a string property called NewLine. What you put into this value will be appended to all strings sent with *WriteLine()* method. It can be "\r\n" or anything appropriate for a device connected to other end of RS-232 link.

In a few places the issue of encoding was mentioned. It applies to a problem of transmission of text data. Many years ago American Standard Code for Information Interchange was created. It took set of characters and assigned them numerical values from 0 to 127, using 7 bits. This way we can store characters in computer memory as bytes of specific values. For example capital letter 'A' has value 65 and lower-case letter 'a' has value 97. Thus string 'Abc' can be stored in memory as an array of following bytes: 65, 98, and 99.

ASCII defines codes for many different characters, including digits. It may come as a surprise but digits have different codes than their values. '0' is encoded as 48, '1' as 49 and so on. Some novice programmers have problems with this fact. It is quite simple though. If you have a byte which stores a number you can convert it to a numerical character by simply adding 48. Of course this applies only to values from 0 to 9.

At this point you may ask what about values such as 3 or 9? What characters are associated with them? Values from 0 to 31 are assigned to control characters, called also non-printable. This comes from the fact that they are not actual graphical characters but rather instructions. We already discussed two such characters: <CR> and <LF>. The other one you are probably familiar with is null-character which is widely used as end of line marker. Generally control characters come from the times when there were many mechanical devices, e.g. teletypewriters. An interesting example is DEL character which has value of 127. This number when written in binary has all 7 bits set to one. Long time ago when paper tape was used to store data, information was written by punching that tape. If there were some holes resembling particular value, DEL could be used to punch all 7 places effectively deleting previously stored value. This also explains why DEL has value of 127, not something from the set 0 – 31.

OK, we have nice set of values for 128 characters. Is it enough? There are many languages in the world and they need own characters which are not present in Latin alphabet. Moreover byte usually consists of 8 bits so why not to use this additional bit and extend ASCII table by 128 more characters? This is what was done. Values from 128 to 255 were used for many different characters, not only for international ones but also for other things. An example are characters used for semi-graphic in text mode. If you were working with text mode programs you probably have seen windows displayed in this mode. Some characters were used for horizontal and vertical lines, some for corners.

Extension of ASCII table solved some problems but also created new ones. There were still thousands of characters which couldn't be stored in memory using just 8 bites for each. Moreover everyone had own idea of what characters should go to extended ASCII table. This was solved by so called code pages. The basic idea of ASCII is that we interpret some numerical value as a character displayed on a screen or printed on paper. With code page we know how to interpret particular value. For example if the code page is ISO-8859-2 than we have to treat values as representing Eastern and Central European characters, if the code page is ISO-8859-1 then we deal with Western European characters.

Unfortunately the situation didn't get better. Many incompatible code pages were created, having different values for the same characters and/or consisting only of subset of characters available in other code page. Dealing with code pages required sending information about used code page. It happened that this information was not available or support for the code page wasn't available on some system or in application. And still there was the problem of limited number of available characters.

To provide sufficient set of characters the Unicode standard was introduced. It offers more than 100 000 characters. Thanks to this we don't need code pages because each value can be unambiguously decoded. However there is still an issue of storing Unicode codes in memory, having only 8-bit bytes. This to some extent was solved with new encoding schemes. The most popular one is UTF-8 which stores ASCII characters the same way as that old standard. This means that ASCII compatible application can read all characters from UTF-8 file as long as they are included in ACII table. For characters of codes greater than 127 the UTF-8 encoding uses 2 to 4 bytes. This way it's variable-length encoding. Java, .NET and Windows operating system use UTF-16 which generally stores characters using 2 bytes. Some characters are encoded using 4 bytes (surrogate pair). In .NET a char variable takes 2 bytes so in case of surrogate pair it has to be stored in two char values. However if you don't have to work with Chinese or other exotic characters you don't need to care about surrogate pairs. Unicode characters can be stored using 4 bytes and it is then called UTF-32. In this standard 32 bits just store the code of a Unicode character. In other words we can say that UTF-32 is nothing more than XXL variant of ASCII. This standard is rarely used though. There also other encodings, like UTF-7 being a version of UTF-8 for 7-bit environments, but they are not very popular.

As you can see, text data may be stored and transmitted in many ways. If you wish to send and receive such data you have to take care of encoding. You can write own routines to handle this task or use encoding schemes provided by *SerialPort* class. It has a property called Encoding which can have several values:

- ASCIIEncoding (default) – for ASCII characters, having codes from 0 to 127

- UTF7Encoding – for UTF-7

- UTF8Encoding – for UTF-8

- UnicodeEncoding – for UTF-16

- UTF32Encoding – for UTF-32

Our intention is to transmit data using serial port. But which one? Computers rarely have more than two serial ports but even if there is only one it would be good to start with enumeration of available ports. Of course for testing purposes you can hardcode the name of the used port (usually COM1) but for final application, designed for use by other people, possibility of selecting desired port would be quite important. Enumeration of available ports is very easy an can be done with simple single line:

```
string[] theSerialPorts =
System.IO.Ports.SerialPort.GetPortNames();
```

As you can see we use *GetPortNames()* static method from *SerialPort* class located in *System.IO.Ports* namespace. It returns an array of string values which we can print inside console window using *foreach* loop.

It's time to make some RS-232 transmission eventually. It will be very simple because no additional hardware is needed. How can we send and receive data not plugging anything to our serial port? Great simplicity of RS-232 allows us to do something which would be not possible with for example USB. As you remember, serial port has TxD and RxD pins. Why not to just connect them with a wire? This way we create a loopback, transmitted data are immediately received. Maybe it doesn't make much sense but it's trivial and is great for learning. So take a look at serial port socket in your computer or USB adapter, locate pins 2 and 3 and connect them. You can use a short cable with crimp pins. It's good to put them inside plastic connector housings to avoid short circuit with other pins of serial port. Alternative solution is to take a jumper connector used for connecting pins, e.g. in IDE hard disks or some motherboards and other electronic devices.

When hardware is ready it's time to write some simple application. You can open Visual Studio and create new C# project (Console Application) or simply create a text file with .cs extension. This file can be compiled using *csc.exe* (C# compiler). It is usually located in a subfolder of *C:\Windows\Microsoft.NET\Framework*. The name of this subfolder depends on version of installed .NET Framework. SerialPort class was introduced in .NET 2.0 so you have to choose a version not lower that this one. For example it can be version 3.5 so the command line to compile source code saved as *serial.cs* file on the desktop can look like this (in one line):

```
C:\Windows\Microsoft.NET\Framework\v3.5\csc.exe
Desktop\serial.cs
```

Here it was assumed that current directory is user's profile directory. The whole path is given here because by default the directory where csc.exe is stored is not included in *PATH* environment variable. After you add the location of csc.exe to PATH variable you can issue just *csc* command in the directory where source file is located.

Having compilation talked over let's look at sample program:

```csharp
using System;
using System.IO.Ports;

namespace SerialConsole
{
    class Program
    {

        static void Main(string[] args)
        {
            string[] serialPorts =
System.IO.Ports.SerialPort.GetPortNames();
            foreach (string port in serialPorts) {
                Console.WriteLine(port);
            }
            Console.Write("Enter name of serial port you want
to use: ");
            string com = Console.ReadLine();
            SerialPort ComPort = new SerialPort(com);
            ComPort.BaudRate = 57600;
            ComPort.Parity = Parity.None;
            ComPort.StopBits = StopBits.One;
            ComPort.DataBits = 8;
            ComPort.Handshake = Handshake.None;
            ComPort.DataReceived += new
SerialDataReceivedEventHandler(OnSerialDataReceived);
            ComPort.Open();
            ComPort.WriteLine("Hello World!");
            Console.ReadKey();
            ComPort.Close();
        }

        private static void OnSerialDataReceived(object
sender, SerialDataReceivedEventArgs args)
        {
            SerialPort ComPort = (SerialPort) sender;
            Console.WriteLine(ComPort.ReadLine());
        }
```

```
    }
}
```

The first referenced namespace allows us to use Console object so we can output some text on a screen. The second one gives access to SerialPort class. The program begins with enumeration of available serial ports. We can iterate over them and display so user can choose a serial port to use. The name is not case sensitive so you can type either „COM3" or „com3" (without quotation marks) and it would work if such port is available. When serial port object is ready, it is being configured. As long as hardware handshake is not used it doesn't really matter what the settings are here because we use a loopback. Our computer is both a transmitter and receiver so settings would match anyway. When transmission options are set then event handler is registered. As you probably noticed it is initiated in a different way than it was previously shown. It comes from the fact that Main function is a static one so the handler has to be static too. This requires creation of *SerialDataReceivedEventHandler* object. When properties of *ComPort* object are ready, it's time to open the port. From this moment we can use it to send data and the handler will be called upon data arrival. And now comes the most interesting part: sending data over RS-232. *WriteLine()* method sends "Hello World!" string with a new line marker. When the *Main()* function waits for a key to be pressed the data go back to the serial port and the incoming data handler is invoked. We instantiate here new SerialPort object but it's initialized with existing one thanks to sender object. We could make *ComPort* global and access it directly from the handler but it won't work when functions are static. When serial port object is obtained we are ready to read received data. It is done with *ReadLine()* method and the data are displayed on a screen. After a key on a keyboard is pressed, the serial port is closed and the program terminates.

Remark

In this example we are passing line of text through loopback. We are sending with *WriteLine()* method and receiving with *ReadLine()*. If you are going to experiment with different hardware configuration, e.g. a device which sends some data not terminated with new line characters, change *ReadLine()* to *ReadExisting()*. In other case you are going to get an exception when the *Close()* method of *SerialPort* object is called but *ReadLine()* still waits for termination of the line.

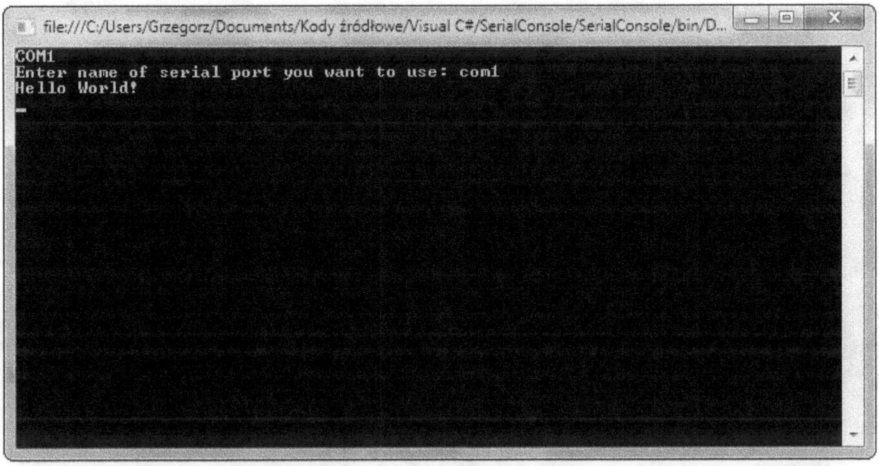

Fig. 5.1.1. Console of example application.

Our program is working properly and gives us some practical knowledge. More information you can find in following chapters, where more real life applications are presented. But there is one more thing you should be aware of – writing code for graphical programs looks different is some details. Thus we have to analyze one more piece of sample code. It's a basic Windows Forms application. This is not the whole source code because the part responsible for creating the main window was created automatically by Visual Studio and is not important now. Another remark: the part of code responsible for choice of serial port was removed as it is not a subject of our consideration now and the name of a serial port was simply hardcoded.

```
using System.Windows.Forms;
using System.IO.Ports;

namespace SerialTest
{
    public partial class Form1 : Form
    {
        SerialPort ComPort = new SerialPort("COM3");
        public Form1()
        {
            InitializeComponent();
            if (ComPort.IsOpen) ComPort.Close();
            ComPort.BaudRate = 57600;
            ComPort.Parity = Parity.None;
            ComPort.StopBits = StopBits.One;
            ComPort.DataBits = 8;
            ComPort.Handshake = Handshake.None;
            ComPort.DataReceived += OnSerialDataReceived;
            ComPort.Open();
            ComPort.WriteLine("Hello World!");
        }

        private void OnSerialDataReceived(object sender,
SerialDataReceivedEventArgs args)
        {
            string txt = ComPort.ReadLine();
            MessageBox.Show(txt);
        }
    }
}
```

As you can see the code is very similar to the previous one, designed for running in console mode. The difference lies in the fact that we make use of a constructor of window's class. This function is not static so we don't have to make data handler static too and setting of DataReceived property of *ComPort* object is a little bit easier. We as well can make *ComPort* object global, at least in a scope of the class.

OK, so for now everything looks quite easy and we are ready to write Windows Forms applications able to communicate over RS-232. Is that true? To some extent the answer is yes, but sometimes things can be tricky. Presenting received data in a message box may be appropriate for some purposes but often we would like to have them shown in a text box, label, or other way. So let's add a text box to our window. If you add a text box to the form it will get a default name textBox1. Let's stick with it and try to write the code for it. Actually it's trivial, just change

```
MessageBox.Show(txt);
```

to

```
textBox1.Text = txt;
```

Are you able to compile the project? There should be no problems. But can you run it? If you are lucky then the program would run correctly and you will see "Hello World!" in a text box. But most probably you will receive following exception: *System.InvalidOperationException*. Why? Additional information says that there was a try to access the text box from a thread which didn't create it. Yes, that's the actual cause. There is a thread in our application responsible for GUI (the window) and a thread handling incoming data. We can't directly access GUI element such as a text box from a different thread. What can we do to circumvent this problem? The answer is very simple. GUI elements have a method called Invoke() and we can use it to access their properties from different threads. Following example shows implementation details.

Serial port and microcontrollers

```csharp
using System.Windows.Forms;
using System.IO.Ports;
using System;

namespace SerialTest
{
    public partial class Form1 : Form
    {
        SerialPort ComPort = new SerialPort("COM3");
        string txt = "";
        public Form1()
        {
            InitializeComponent();
            if (ComPort.IsOpen) ComPort.Close();
            ComPort.BaudRate = 57600;
            ComPort.Parity = Parity.None;
            ComPort.StopBits = StopBits.One;
            ComPort.DataBits = 8;
            ComPort.Handshake = Handshake.None;
            ComPort.DataReceived += OnSerialDataReceived;
            ComPort.Open();
            ComPort.WriteLine("Hello World!");
        }

        private void OnSerialDataReceived(object sender,
SerialDataReceivedEventArgs args)
        {
            txt = ComPort.ReadLine();
            textBox1.Invoke(new
EventHandler(textBox1_DataReceived));
        }

        private void textBox1_DataReceived(object sender,
EventArgs e)
        {
            textBox1.Text = txt;
        }
    }
}
```

As we can see a new method was added. It makes actual operation on GUI element and is specified as a parameter of the constructor of an object of *EventHandler* class. This object is a parameter for *Invoke()* method of the text box. The declaration of txt variable was moved outside of *OnSerialDataReceived()* function. Another change is the added declaration of System namespace where the *EventHandler* class is located.

Remark

When you write software using *SerialPort* class you can sometimes get following error:

System.IO.IOException: The I/O operation has been aborted because of either a thread exit or an application request.

This error doesn't happen very often but when it occurs it's pretty nasty. Of course it can be programmer's fault but it may be also caused by a bug which exists in Microsoft .NET. The details about this bug and a workaround can be found on the following website: http://zachsaw.blogspot.com/2010/07/net-serialport-woes.html

5.2. Programming serial port in C/C++ with Windows API

Although C# makes writing software easy, there can be situations where .NET is not welcome and there is need for pure WinAPI code. API of Windows operating system of course supports serial port but using it is not as pleasurable as writing in .NET. Anyway it's possible and can be not so stressful, especially if you design own wrapper classes or functions.

For accessing serial ports Windows API provides the same functions as for working with files: *CreateFile()* for opening a port, *WriteFile()* for sending data, *ReadFile()* for receiving data, and *CloseHandle()* for closing a port. Let's see how they are supposed to be used for serial communication.

```
HANDLE WINAPI CreateFile(
   __in       LPCTSTR lpFileName,
   __in       DWORD dwDesiredAccess,
   __in       DWORD dwShareMode,
   __in_opt   LPSECURITY_ATTRIBUTES lpSecurityAttributes,
   __in       DWORD dwCreationDisposition,
   __in       DWORD dwFlagsAndAttributes,
   __in_opt   HANDLE hTemplateFile
);
```

Despite its name, *CreateFile()* is not only for creating new files but also for opening existing ones. Moreover it works not only with disk files but with many I/O objects, including serial ports. To open a serial port we have to provide a port name as first parameter, for example "COM5". The *dwDesiredAccess* parameter determines access mode. Usually we want to have both sending and receiving enabled so we use combination of flags *GENERIC_READ* and *GENERIC_WRITE*. *dwShareMode* determines sharing mode, in case of serial ports it must be zero. We don't specify any security attributes as well so the pointer to *SECURITY_ATTRIBUTES* structure should be NULL. As for the *dwCreationDisposition* parameter we must specify *OPEN_EXISTING*.

Now comes a tricky part, the *dwFlagsAndAttributes* parameter. Here can use the *FILE_FLAG_OVERLAPPED* flag. What does it do? It enables so called overlapped I/O operations. Without overlapping a program waits until an input/output operation finishes. Until completion of the operation the thread which called the operation is blocked. In practice this means that when we call *ReadFile()* the execution of our program will be stopped until some data are received. Similar condition applies to *WriteFile()* but this is usually not big deal because sending data takes short time while waiting for data to come can take very long time. To solve this problem Microsoft introduced overlapped I/O operations. The idea is that calls to *ReadFile()* and *WriteFile()* don't block the thread because they just initiate input/output operations. The program can continue with execution of other tasks while the I/O operations are performed in background. When they are finished we can get their results. The subject of overlapped I/O will be presented in details later. As for the *dwFlagsAndAttributes* parameter you need to remember that you have to turn on the *FILE_FLAG_OVERLAPPED* flag to enable overlapping or set the parameter to 0 to disable overlapping. The *hTemplateFile* parameter should be set to NULL. The *CreateFile()* function returns handle to opened serial port (file in general). In case of error the *INVALID_HANDLE_VALUE* value is returned.

```
BOOL WINAPI WriteFile(
    __in         HANDLE hFile,
    __in         LPCVOID lpBuffer,
    __in         DWORD nNumberOfBytesToWrite,
    __out_opt    LPDWORD lpNumberOfBytesWritten,
    __inout_opt  LPOVERLAPPED lpOverlapped
);
```

To send data through a serial port we can use the *WriteFile()* function. The first parameter is a handle of a serial port opened previously using *CreateFile()*. The second parameter points to a buffer containing bytes to be sent and the third one tells how many of these bytes should be sent. Naturally size of the buffer must be at least as big as the number of bytes we want to send. When the *WriteFile()* function finishes it stores number of bytes it managed to send in a variable pointed by *lpNumberOfBytesWritten*. The last parameter is a pointer to the *OVERLAPPED* structure which we use for overlapped I/O operations. If we don't use overlapping this parameter must be NULL.

To close handle of opened port we use the *CloseHandle()* function, just like for regular files.

At this moment we know how to open a serial port and transmit data. What else we need is to configure transmission parameters, like baudrate, parity or stop bits. For all these settings Windows API uses one big structure called *DCB*. Its definition is as follows:

```
typedef struct _DCB {
  DWORD DCBlength;
  DWORD BaudRate;
  DWORD fBinary   :1;
  DWORD fParity   :1;
  DWORD fOutxCtsFlow   :1;
  DWORD fOutxDsrFlow   :1;
  DWORD fDtrControl   :2;
  DWORD fDsrSensitivity   :1;
  DWORD fTXContinueOnXoff   :1;
  DWORD fOutX   :1;
  DWORD fInX   :1;
  DWORD fErrorChar   :1;
  DWORD fNull   :1;
  DWORD fRtsControl   :2;
  DWORD fAbortOnError   :1;
  DWORD fDummy2   :17;
  WORD  wReserved;
  WORD  XonLim;
  WORD  XoffLim;
  BYTE  ByteSize;
  BYTE  Parity;
  BYTE  StopBits;
```

```
    char   XonChar;
    char   XoffChar;
    char   ErrorChar;
    char   EofChar;
    char   EvtChar;
    WORD   wReserved1;
} DCB, *LPDCB;
```

Let's briefly explain some of more important fields available in this structure. *DCBlength* is size of the structure and should always be set using *sizeof(DCB)*. *BaudRate* sets speed of transmission in bits per second. It can be actual number or one of predefined constants, from *CBR_110* to *CBR_256000*. *fBinary* turns on binary transmission. As Windows doesn't support nonbinary modes this member of structure must be set to *TRUE*. fParity enables/disables parity checking, can be set to *TRUE* or *FALSE*. If *fOutxCtsFlow* set to *TRUE* the CTS (clear-to-send) signal is monitored for output flow control. If this fields is *TRUE* and CTS is turned off, output is suspended until CTS is sent again. Similar situation applies to *fOutxDsrFlow* member of *DCB* structure. This field controls monitoring of DSR (data-set-ready) signal. In case of *fDtrControl* field there are three values possible: 0 (*DTR_CONTROL_DISABLE*) - disables the DTR line when the device is opened and leaves it disabled, 1 (*DTR_CONTROL_ENABLE*) - enables the DTR line when the device is opened and leaves it on, and 2 (*DTR_CONTROL_HANDSHAKE*) – enables DTR handshaking. If *fDsrSensitivity* is set to TRUE, serial port ignores any bytes received, unless the DSR modem input line is high. The *fRtsControl* field can have several values: 0 (*RTS_CONTROL_DISABLE*) – disables the RTS line when the serial port is opened and leaves it disabled, 1 (*RTS_CONTROL_ENABLE*) – enables the RTS line when the serial port is opened and leaves it on, 2 (*RTS_CONTROL_HANDSHAKE*) – enables RTS handshaking, 3 (*RTS_CONTROL_TOGGLE*) – the RTS line is high during transmission and low when the transmission is finished. *ByteSize* determines how many bits are in one transmitted/received frame of data, usually 8. While *fParity* member field enables or disables parity checking, the *Parity* field specifies type of parity check. Possible values are: 0 (*NOPARITY*), 1 (*ODDPARITY*), 2 (*EVENPARITY*), 3 (*MARKPARITY*), 4 (*SPACEPARITY*). The *StopBits* field sets number of stop bits: 0 (*ONESTOPBIT*) – for one stop bit, 1 (*ONE5STOPBITS*) – for 1.5 stop bits, and 2 (*TWOSTOPBITS*) – for 2 stop bits.

There are three functions which operate on the *DCB* structure:

```
BOOL WINAPI GetCommState(HANDLE hFile, LPDCB lpDCB);
```
– fills the structure with current settings

```
BOOL WINAPI SetCommState(HANDLE hFile, LPDCB lpDCB);
```
– sets serial port options using the *DCB* structure

```
BOOL WINAPI BuildCommDCB(LPCTSTR lpDef, LPDCB lpDCB);
```
– allows to set certain fields in *DCB* using a string value

Usage of the first two functions is straightforward, the first parameter for both of them is a handle of currently opened serial port and the second one is a pointer to *DCB* structure. In case of *BuildCommDCB()* we need to provide a string containing transmission parameters. Its syntax is as follows:

```
[baud=b][parity=p][data=d][stop=s][to={on|off}][xon={on|off}]
[odsr={on|off}][octs={on|off}][dtr={on|off|hs}][rts={on|off|h
s|tg}][idsr={on|off}]
```

For example, for 9600 bits per second baud rate, 8 bits of data, 1 stop bit and no parity we should use following string:

```
"baud=9600 parity=N data=8 stop=1"
```

As you can see, Microsoft describes the type of string as LPCTSTR, which stands for Long Pointer to a Const TCHAR STRing. It means we can use regular, null-terminated C-style string. But it may happen that you will be writing a Unicode application and you will have the UNICODE macro defined in your program. In this case

BuildCommDCB is redefined as *BuildCommDCBW* which has first parameter of type *LPCWSTR*. It is similar to *LPCTSTR* but for so called wide characters (Unicode). You have to be aware of this, especially if you have a string obtained from other function. If you are going to use a string literal, a good practice is to use the _T() macro which returns a regular or wide-character string depending whether we are compiling Unicode application or not. Presented remarks actually apply to many Windows API functions which are available in two forms: ANSI (suffix A) and Unicode (suffix W).

```
#ifdef UNICODE
#define BuildCommDCB   BuildCommDCBW
#else
#define BuildCommDCB   BuildCommDCBA
#endif // !UNICODE
```

Let's examine simple example showing how to configure serial port and then read and write data. Here we are not using overlapping, I/O function are blocking execution of the program. The program is meant to work with a device working as described in one of examples presented in this book. There is a microcontroller with a button and a LED. The LED can be turned on by sending 'a' and turned off by sending 'b'. When the button is pressed the microcontroller sends 'P' and 'N' when released. Here is the code:

```
#include <tchar.h>
#include <windows.h>
#include <iostream>
#include <string>

using namespace std;

int _tmain(int argc, _TCHAR* argv[])
{
      HANDLE m_hCommPort = CreateFile(_T("COM1"),
GENERIC_READ | GENERIC_WRITE, 0, NULL, OPEN_EXISTING, NULL,
NULL);
      if (m_hCommPort == INVALID_HANDLE_VALUE) {
```

```
                cerr << "Error: " << GetLastError() << endl;
                return 1;
        }

        DCB dcb;
        memset(&dcb, 0, sizeof(dcb));
        dcb.DCBlength = sizeof(dcb);
        if (!BuildCommDCB(_T("baud=9600 parity=N data=8
stop=1"), &dcb))
        {
                cerr << "Error: " << GetLastError() << endl;
                return 1;
        }
        dcb.fOutxCtsFlow = FALSE;
        dcb.fOutxDsrFlow = FALSE;
        dcb.fDtrControl = DTR_CONTROL_DISABLE;
        dcb.fDsrSensitivity = FALSE;
        dcb.fOutX = FALSE;
        dcb.fInX = FALSE;
        dcb.fNull = FALSE;
        dcb.fRtsControl = RTS_CONTROL_DISABLE;
        dcb.fAbortOnError = FALSE;
        if (!SetCommState(m_hCommPort,&dcb))
        {
                cerr << "Error: " << GetLastError() << endl;
                return 1;
        }

        while(1) {
                string i;
                cin >> i;
                char buf[1];
                if (i.compare("on")==0) buf[0] = 'a';
                if (i.compare("off")==0) buf[0] = 'b';
                if (i.compare("q")==0) {
                        CloseHandle(m_hCommPort);
                        return 0;
                }

                DWORD written;
                WriteFile(m_hCommPort, buf, 1, &written, NULL);

                DWORD read;
```

```
              ReadFile(m_hCommPort, buf, 1, &read, NULL);
              if (buf[0]=='P') cout << "Button pressed." <<
endl;
                    else cout << "Button released." << endl;
       }

       return 0;
}
```

The program uses several header files. *tchar.h* provides *_T()* macro, *windows.h* gives access to Windows API functions and structures, *iostream* and *string* are standard C++ libraries giving access to C++ streams and strings.

On the beginning we are opening serial port, here it is COM1, without overlapping functionality. When the port is opened we are preparing the *DCB* structure to set transmission parameters. First we are zeroing whole structure using *memset()*, it's an easy way to initialize many fields, including those which have to be set to zero, e.g. *wReserved*. The next step is to set DCBlength to size of whole structure, an action which is required for many structures used by Windows API functions. Then some transmission parameters are set with BuildCommDCB() and some with direct write to *DCB* structure fields. Finally settings for the opened port are changed using *SetCommState()* function.

When the serial port is opened and configured the most interesting part comes. The algorithm is simple: we are reading a command from standard input and sending appropriate byte through RS-232 or exiting. When the byte is sent, we are waiting for a byte to come. After you start the program you need to type something. If you type "on" the program will send the 'a' character (byte of value 97), if you type "off" the program will send the 'b' character (byte of value 98). If you type "q" the program will close the serial port and terminate without sending anything. After a byte is sent the application waits for one byte to come from RS-232 link. You have to pay attention to that. It means that if no data were received the program will wait.

But if some data were received when program was waiting for you to enter a command, the *ReadFile()* function will return immediately storing the first received byte in provided buffer. If there were more bytes received they will be acquired during subsequent calls to *ReadFile()*.

It's really good to experiment with mentioned circuit consisting of a microcontroller, button and LED. You can observe then following behavior: you press a button, release it and then enter some command, for example "on". The program will tell you that the button is pressed. You enter some command again (different that "q") without pressing the button and the program tells you that the button is released. Do you see the problem? For some time you see incorrect information, you are not informed about releasing a button until you entered another command and the program had a chance to read another byte. As you see the *ReadFile()* function reads bytes from serial port's input buffer. It tries to read as many bytes as a value of the third parameter it was called with. If there are as many bytes as the parameter says, the function copies them into provided buffer. If there are less bytes the function waits for more to come. You can check it by changing the parameter from 1 to 2 and of course by resizing the *buf* buffer accordingly (change `char buf[1];` to `char buf[2];`). The piece of code responsible for receiving data can look as follows:

```
ReadFile(m_hCommPort, buf, 2, &read, NULL);

for (int i = 0; i< read; i++) {
        if (buf[i]=='P') cout << "Button pressed." << endl;
            else cout << "Button released." << endl;
}
```

Start the program and enter some command, for example "on". Now press the button – nothing happens. Release the button – program shows information about pressing and releasing of the button. As you can see the program waits until two bytes are received.

172

It's clearly visible that presented way of reading and writing data to serial port is not very flexible. It's good enough if we know how much data are going to come and we don't need to do anything while waiting for incoming bytes. But what in other situations, where we want to send and receive data asynchronously and regardless of their size? There are two things which we can use, separately or in a conjunction. They are: overlapped I/O operations and threads.

What are overlapped operations? If a serial port is opened in overlapped mode then functions such as *WriteFile()* or *ReadFile()* don't perform whole operation from start to finish but only initiate it. The I/O operation is executed in background and the program may perform other tasks. When those tasks are finished results of the I/O operation can be retrieved. To work with overlapped operations we need two things: an event object and the *OVERLAPPED* structure. Event objects generally are synchronization objects which are useful in sending a signal to a thread indicating that a particular event has occurred. In serial communication we use them to obtain information about communication events, such as new data arrival. The *OVERLAPPED* structure contains information about particular asynchronous I/O operation, including a handle to event object that will be set to a signaled state by the system when the operation has completed.

You can use following functions to work with overlapped I/O operations:

- `HANDLE WINAPI CreateEvent(LPSECURITY_ATTRIBUTES lpEventAttributes, BOOL bManualReset, BOOL bInitialState, LPCTSTR lpName);` - creates an event object which we can use for overlapped operation

- `BOOL HasOverlappedIoCompleted(LPOVERLAPPED lpOverlapped);` - it's actually a macro which checks whether an event pointed by OVERLAPPED structure stopped being in pending state and hence the I/O operation is finished

- `DWORD WINAPI WaitForSingleObject(HANDLE hHandle, DWORD dwMilliseconds);` - waits for an event object to be in signaled state or a time-out occurs

- `BOOL WINAPI GetOverlappedResult(HANDLE hFile, LPOVERLAPPED lpOverlapped, LPDWORD lpNumberOfBytesTransferred, BOOL bWait);` - retrieves result of an overlapped operation: whether is succeeded or not and how many bytes were transferred

Generally the flow is as follows:

1. Create an event object using *CreateEvent()*

2. Initialize OVERLAPPED structure with a handle to created event object

3. Call *WriteFile()* or *ReadFile()* with OVERLAPPED structure as the last parameter

4. Check if the operation has already finished by calling HasOverlappedIoCompleted()

5. If the operation is not finished yet wait some time using WaitForSingleObject()

6. When the operation is complete call *GetOverlappedResult()*

This is just an example algorithm; you can use a bit different solution. Particularly you don't need to use *HasOverlappedIoCompleted()* or *WaitForSingleObject()* functions, you may choose to use Sleep() function for delay and check value returned by *GetOverlappedResult()*. If *GetOverlappedResult()* fails it returns zero, if it succeeds the return value is nonzero.

As it was said earlier, you can use threads to perform some tasks during waiting for incoming data. This way the code responsible for data transmission can run independently of the rest of your program. This is a neat solution but besides of synchronization problems typical for threading application you still have to pay attention whether there are blocking or overlapped I/O operations used. Even if you use threads the serial port operations are serialized when you opened the port in blocking mode. It means that if one thread waits for data the other can't send anything until the read operation is finished. Thus you need to use overlapping in threading programs too. Of course it doesn't apply if you don't need asynchronous reads and writes, if you call them sequentially then blocking should not pose any problem.

Here is a simple example showing how to use overlapped I/O and threading.

```
#include <tchar.h>
#include <windows.h>
#include <iostream>
#include <string>

using namespace std;

bool newCommand = false;
string i;

BOOL initSerial(HANDLE hCommPort);
DWORD WINAPI inputThread(LPVOID lpParameter);

BOOL initSerial(HANDLE hCommPort) {
        DCB dcb;
        memset(&dcb,0,sizeof(dcb));
        dcb.DCBlength=sizeof(dcb);
        if (!BuildCommDCB(_T("baud=9600 parity=N data=8
stop=1"),&dcb))
        {
                return false;
        }

        // Common settings
        dcb.fOutxCtsFlow = FALSE;
```

```
        dcb.fOutxDsrFlow = FALSE;
        dcb.fDtrControl = DTR_CONTROL_DISABLE;
        dcb.fDsrSensitivity = FALSE;
        dcb.fOutX = FALSE;
        dcb.fInX = FALSE;
        dcb.fNull = FALSE;
        dcb.fRtsControl = RTS_CONTROL_DISABLE;
        dcb.fAbortOnError = FALSE;
        return SetCommState(hCommPort,&dcb);
}

int _tmain(int argc, _TCHAR* argv[])
{
        HANDLE hCommPort = CreateFile(_T("COM1"), GENERIC_READ
| GENERIC_WRITE, 0, NULL, OPEN_EXISTING,
        FILE_FLAG_OVERLAPPED, NULL);
        if (hCommPort == INVALID_HANDLE_VALUE) return 1;

        if (!initSerial(hCommPort)) return 1;

        HANDLE hThread = CreateThread(NULL, 0, inputThread,
NULL, 0, NULL);

        DWORD dwRead;
        char buf[1];
        OVERLAPPED osReader = {0};
        osReader.hEvent = CreateEvent(NULL, TRUE, FALSE, NULL);

        bool running = true;
        while(running) {

                ReadFile(hCommPort, buf, 1, &dwRead, &osReader);
                while(!HasOverlappedIoCompleted(&osReader) ||
!running) {
                        if (newCommand) {
                                newCommand = false;
                                char buf[]={'a'};
                                if (i.compare("on")==0) buf[0] =
'a';
                                if (i.compare("off")==0) buf[0] =
'b';
                                if (i.compare("q")==0) {
                                        running = false;
```

```
                        break;
                    }
                    DWORD written;
                    OVERLAPPED osWriter = {0};
                    osReader.hEvent = CreateEvent(NULL,
TRUE, FALSE, NULL);

                    WriteFile(hCommPort, buf, 1,
&written, &osWriter);

        WaitForSingleObject(osWriter.hEvent, 100);

                    }
                }

            if (running)
                    if (GetOverlappedResult(hCommPort,
&osReader, &dwRead, FALSE))
                        cout << buf[0] << endl;

        }
        CloseHandle(hCommPort);
        cout << "Exiting." << endl;
}

DWORD WINAPI inputThread(LPVOID lpParameter){
        while(i.compare("q")!=0){
            cin >> i;
            newCommand = true;
        }
        return 0;
}
```

First, we have the *initSerial()* function which sets parameters of serial communication. Details of this code were discussed earlier, here it is just moved to separate function. When a serial port is opened and configured in *_tmain()*, we are creating a new thread which runs the *inputThread()* function. It's a helper function for reading commands from keyboard. It runs in a separate thread because reading standard input is a blocking operation and we want to receive data from serial port while read is pending. If a command is read from keyboard then the *newCommand* variable is set to true. If that command was *q* then the *while* loop finishes and the thread is terminated as well.

When the keyboard reading thread has been created, we start asynchronous read operation for serial port. Then we check if it has finished. If it has, we check the result. If it hasn't we check for new command entered by keyboard. Here we benefit from the overlapped I/O function: we can process commands from keyboard while I/O operation is being executed.

If a command was entered and it is not *q* we send appropriate byte through serial port. The *WriteFile()* function operates in overlapped mode, thus we use *WaitForSingleObject()* to wait for it to finish. Used timeout of 100 milliseconds is short because sending one byte takes very little amount of time. If the command *q* was entered then the helper thread is already terminated. Following this, the main loop terminates, the serial port is closed and execution of the program ends.

Fig. 5.2.1. Console of example application with overlapping.

Overlapped I/O operations help us write code without threads. Here we use a thread just because of blocking keyboard read. Reading data from serial port no longer halts execution of the rest of the program. However if you plan to use threads then be careful. You need to take care about proper synchronization and resources (e.g. memory) management. When the main thread is terminated all the other threads are terminated too and the operating system would free up all used resources, including the serial port. But it's not a good practice, nor is it elegant. Thus we should take care about closing handles and terminating threads. Besides of just finishing main thread you may want to use the *TerminateThread()* function. It can seem to be a simple way of killing a thread but it can cause some side effects, like not releasing a memory the killed thread allocated. In our simple example there is no problem because the thread terminates itself but in other applications you should take care about proper threads management.

5.3. *Programming serial port in C/C++ on Linux*

Windows is very popular operating system but not everyone likes it, it's not free and is not suitable for some purposes. Thus people quite often choose Linux and it would be good to know how to write programs which can communicate over serial port.

Under Linux operating system serial ports are visible as *ttyS** files in */dev* directory, where */dev/ttyS0* is the first serial port, */dev/ttyS1* is the second one and so on. Usually there are more such files than actual physical ports and their number is determined by *CONFIG_SERIAL_8250_RUNTIME_UARTS* setting in kernel configuration. If you want to know which files correspond to available serial ports and which not you can issue following command:

```
ls /sys/class/tty/ttyS*/device/id
```

It makes use of the fact that directories for unused *ttyS** files in */sys/class* don't contain *id* files. The other possibility is to check *dmesg* entries:

```
dmesg | grep ttyS
```

In case of virtual serial ports created by USB adapters the situation is much simpler. They are represented by */dev/ttyUSB** files which are visible only when particular adapters are connected.

On the Linux platform functions for RS-232 communication are different than on Windows but there are some similarities. You must know that serial I/O under Linux is implemented as part of the terminal I/O capabilities of UNIX. Thus in the so called canonical input processing mode the standard piece of data is line of text. The read operation doesn't finish until full line is received. Moreover if there are some special characters they get interpreted, e.g. backspace erases an already transmitter character. This is a historical issue, having origin in times of teletypewriters (hence TTY abbreviation). If you are going to write a program which operates on received lines of text then canonical mode can be a good choice. But we are going to discuss the non-canonical input processing, were you are receiving raw data.

To open a serial port we are going to use standard *open()* function, we can do it in this way:

```
int fd = open("/dev/ttyUSB0", O_RDWR | O_NOCTTY | O_NDELAY);
```

First parameter is path to particular serial port, in above example it is a virtual serial port created by USB<->serial adapter. The second parameter is a combination of flags:

- O_RDWR – enables reading and writing

- O_NOCTTY – tells that the port's terminal will not be allocated as the calling process's controlling terminal, otherwise our program will be affected by events such as keyboard abort signals

- O_NDELAY – the program doesn't wait until connected device is ready (state of DCD line is not checked)

If the function succeeds it returns file descriptor which is then stored in *fd* variable. When the port is opened we can configure parameters of transmission, we do it with the *termios* structure and *tcsetattr()* function. The *termios* structure has several fields:

- c_cflag – control options

- c_lflag – line options

- c_iflag – input options

- c_oflag – output options

- c_cc – array of control characters

- c_ispeed – input baud (new systems)

- c_ospeed – output baud (new systems)

One of more important fields is *c_cflag*, we can set there following flags:

- CS5 .. CS8 – set number of bits in a frame

- CSTOPB – if set then there are two stop bits, if this flag is not set then there is one stop bit

- CREAD – enables receiver

- PARENB – enables parity bit

- PARODD – if set then odd parity is used, if this flag is not set then even parity is used

- CLOCAL – configures the connection as local, without modem control; otherwise the application will be subject to job control or hangup signals

- CNEW_RTSCTS – enables RTS/CTS flow control, this flag is also called CRTSCTS

As you can see the *c_cflag* field is very important because it configures basic parameters of transmission. The *c_lflag* field is important in canonical mode and we are not going to discuss it here. The similar situation is with the *c_iflag* field, just three flags related to XON/XOFF flow control need our attention:

- IXON – enables XON/XOFF flow control for output queue, transmitting data is hold when the other end can't process received data fast enough

- IXOFF – enables XON/XOFF flow control for input queue, when data are coming to fast then the ASCII DC3 character (pause) is sent, when the receiver is ready again then ASCII DC1 character (resume) is sent

- IXANY – when sending data was suspended by XON/XOFF flow control then any incoming character resumes transmission

We also are not going to describe c_oflag, in most cases you won't need to set any flags there. As for the c_cc field, there are definitions of control characters. But this array doesn't only store characters; there are also two very important variables: VMIN and VTIME. VMIN sets minimum number of characters to read – the read operation will block until given number of bytes is received. VTIME sets timeout for incoming data in tenths of seconds. If VMIN is zero then VTIME is a timeout between any two characters. If VMIN is nonzero then VTIME specifies timeout for first character. Timeouts are ignored in canonical input mode or when the NDELAY flag is set on the file via open() or fcntl().

We need to tell something about setting baudrate. It can be set using c_cflag field but it's not the only place, there are also c_ospeed and c_ispeed fields. Which place is used depends on version of operating system, thus it's better to call cfsetospeed() and cfsetispeed() functions.

When the termios structure is correctly filled we can call tcsetattr() to configure the serial port. This is usually done together with calling tcflush() which clears serial port's input and output buffers.

Reading and writing data is very simple, we are using standard read() and write() functions. They take file descriptor, pointer to data buffer and number of bytes to send/receive.

Serial port and microcontrollers

Now we can take a look at sample source code:

```cpp
#include <sys/types.h>
#include <sys/stat.h>
#include <fcntl.h>
#include <termios.h>
#include <iostream>
#include <string.h>

#define MODEMDEVICE "/dev/ttyUSB0"
#define BAUDRATE B9600

using namespace std;

int main() {
    int fd = open(MODEMDEVICE, O_RDWR | O_NOCTTY | O_NDELAY);
    if (!fd) {
        cerr << "Cannot open port " << MODEMDEVICE << endl;
        return 1;
    }
    struct termios serialOpt;
    bzero(&serialOpt, sizeof(serialOpt));
    serialOpt.c_cflag = CS8 | CLOCAL | CREAD;
    serialOpt.c_iflag = 0;
    serialOpt.c_oflag = 0;
    serialOpt.c_lflag = 0;
    serialOpt.c_cc[VTIME]    = 0;    /* inter-character timer
unused */
    serialOpt.c_cc[VMIN]     = 1;    /* blocking read until 1
character received */
    cfsetispeed(&serialOpt, BAUDRATE);
    cfsetospeed(&serialOpt, BAUDRATE);
    tcflush(fd, TCIFLUSH);
    tcsetattr(fd, TCSANOW, &serialOpt);
    char buf[1];
    while(1) {
        string i;
        cin >> i;
        if (i.compare("on")==0) buf[0] = 'a';
        if (i.compare("off")==0) buf[0] = 'b';
        if (i.compare("q")==0) {
            break;
        }
```

```
        write(fd, buf, 1);
        read(fd, buf, 1);
        cout << buf[0] << endl;
    }
    close(fd);
    return 0;
}
```

In this example we are using virtual serial port from USB<->serial adapter and typical transmission parameters: 9600 bps, 8 bits of data, no parity and one stop bit. We are opening serial port and then configuring it. The *termios* structure is zeroed using *bzer()* and then appropriate flags are set. We are not using timeouts but set blocking to one character. Input and output speeds are set to the same value of 9600 bits per second. Then the serial port is flushed and configured. The program works in the same way as it was presented in the chapter devoted to Windows API programming. Depending on string entered by user it sends 'a' or 'b'. Connected microcontroller turns its LED on or off accordingly. If 'q' is entered then program terminates. When the byte was sent the program waits for one byte. When it was received it is displayed and the program waits for user input again. Notice that received byte is displayed after user entered a command, not in the moment it was received. Received bytes are displayed one by one, after each input from keyboard. When no bytes were received then sent byte is displayed. You can improve the logic of the program but still we are dealing with blocking I/O.

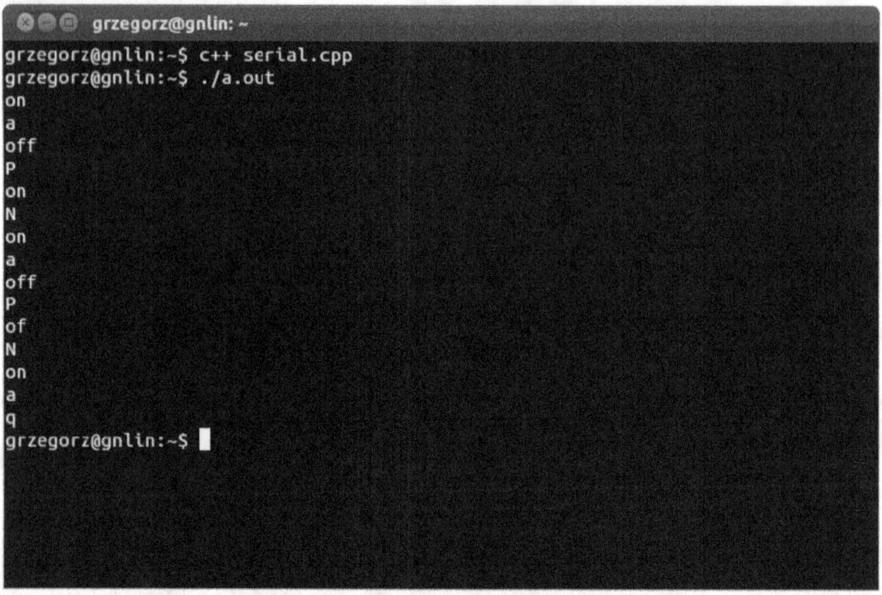

```
grzegorz@gnlin: ~
grzegorz@gnlin:~$ c++ serial.cpp
grzegorz@gnlin:~$ ./a.out
on
a
off
P
on
N
on
a
off
P
of
N
on
a
q
grzegorz@gnlin:~$ 
```

Fig. 5.3.1. Console of the example with blocking I/O.

A different approach is to use signals. It's a form of inter-process communication which Linux implements as a POSIX-compliant operating system. They are kind of messages notifying about some events and interrupting normal program execution. For example, if you press Ctrl+C when a console program is running, then the *SIGINT* signal will be sent and the program will be terminated; in case of segmentation fault the *SIGSEGV* is sent. We can use signals for being notified about incoming data from serial port. Thanks to them we don't need to call blocking read operation so it can wait for data, instead we can call read when we know there are some data received and ready to be acquired. The signal we are interested in is SIGIO. For handling serial port events we need to fill *sigaction* structure where we set pointer to signal handler function, and call *fcntl()* to enable signals for our process, coming from file descriptor belonging to opened serial port. A simple example is shown below:

```cpp
#include <sys/types.h>
#include <sys/stat.h>
#include <fcntl.h>
#include <termios.h>
#include <stdio.h>
#include <sys/signal.h>
#include <unistd.h>
#include <string.h>

#define SERIAL_PORT "/dev/ttyS0"
#define BAUDRATE B9600

using namespace std;

void signal_handler_IO (int status);

int newData = 0;

int main() {
    int fd = open(SERIAL_PORT, O_RDWR | O_NOCTTY);
    if (!fd) {
        cerr << "Cannot open port " << MODEMDEVICE << endl;
        return 1;
    }
    struct sigaction saio;
    saio.sa_handler = signal_handler_IO;
    sigemptyset(&saio.sa_mask);
    saio.sa_flags = 0;
    sigaction(SIGIO, &saio, NULL);
    fcntl(fd, F_SETOWN, getpid());
    fcntl(fd, F_SETFL, FASYNC);

    struct termios newtio;
    bzero(&newtio, sizeof(newtio));
    newtio.c_cflag = BAUDRATE | CS8 | CLOCAL | CREAD;
    newtio.c_iflag = 0;
    newtio.c_oflag = 0;
    newtio.c_lflag = 0;
    newtio.c_cc[VTIME] = 0;
    newtio.c_cc[VMIN] = 1;
    tcflush(fd, TCIFLUSH);
    tcsetattr(fd, TCSANOW, &newtio);
```

```
    char buf[] = {'a'};
    write(fd, buf, 1);
    while(1){
        if(newData) {
            newData = 0;
            read(fd, buf, 1);
            printf("%c\n", buf[0]);
        }
        // some operations
    }
    close(fd);
    return 0;
}

void signal_handler_IO (int status) {
    printf("SIGIO signal received.\n");
    newData = 1;
}
```

When a serial port is opened we configure signals handling. Using *sa_handler* field in *sigaction* structure we define handler function for received signals. There are no flags set and the mask is empty – no other signals are blocked during execution of the signal handler. When the structure is filled we call *sigaction()* to enable signals for our process. The last two things are calls to *fcntl()*, first one to allow the process to receive SIGIO and the second one to make the file descriptor asynchronous.

Configuration of serial port is performed in usual way. When everything is configured the program sends *a* to turn on the LED. Then it checks if the *newData* flag was changed by the signal handler. If it was then then it reads received byte and displays it. Reading and displaying received data could be done on the signal handler routine but such an approach should be avoided although it works.

At the end of the *main()* function opened serial port is closed but this line of code is never reached. The *while* loop doesn't check any condition and you have to press Ctrl+C or terminate the process in other way to close the program. This is just for the example to be simple, in your program you need to add some logic in order to control execution flow and have program terminated in proper way. It can be something like this:

```
while(buf[0]!='N'){
    if(newData) {
        newData = 0;
        read(fd, buf, 1);
        printf("%c\n", buf[0]);
    }
    // some operations
}
```

The program is terminated when the button is released and N is received. Of course everything is up to you, for example you can read keyboard.

There is one more interesting thing you for sure already noticed if you ran the program. You see a message about received *SIGIO* signal even if no byte was received at all. This is because *SIGIO* signals are sent not only for input operations but also for output operations. The signal you see is caused by the write operation our program executes on the beginning. If you remove this line you won't see that signal. So if you want to use *SIGIO* as an indicator of incoming data you need to take care about distinguishing between input and output signals, for example by clearing the flag (here called *newData*) after write operations.

Another approach to asynchronous read and write operations when working with a serial port is to use threads. It's quite easy and here is an example:

```cpp
#include <fcntl.h>
#include <termios.h>
#include <pthread.h>
#include <string.h>
#include <iostream>

#define SERIAL_PORT "/dev/ttyS0"
#define BAUDRATE B9600

using namespace std;

void * readThread(void * threadId);

int fd;

int main() {
    fd = open(SERIAL_PORT, O_RDWR | O_NOCTTY);
    if (!fd) {
        cerr << "Cannot open port " << MODEMDEVICE << endl;
        return 1;
    }
    struct termios newtio;
    bzero(&newtio, sizeof(newtio));
    newtio.c_cflag = BAUDRATE | CS8 | CLOCAL | CREAD;
    ncwtio.c_iflag = 0;
    newtio.c_oflag = 0;
    newtio.c_lflag = 0;
    newtio.c_cc[VTIME] = 0;
    newtio.c_cc[VMIN] = 1;
    tcflush(fd, TCIFLUSH);
    tcsetattr(fd, TCSANOW, &newtio);

    pthread_t thread;
    pthread_create(&thread, NULL, readThread, (void *)0);

    while(1){
        char buf[1];
        string s;
        cin >> s;
        if (s.compare("on")==0) buf[0] = 'a';
        if (s.compare("off")==0) buf[0] = 'b';
        if (s.compare("q")==0) {
            break;
```

```
        }
        write(fd, buf, 1);
    }
    close(fd);
    return 0;
}

void * readThread(void * threadId)
{
    char buf[1];
    while(1) {
        read(fd, buf, 1);
        cout << buf[0] << endl;
    }
}
```

Opening and configuring a serial port is done as previously. When the port is ready we start a new thread using *pthread_create()* where we provide pointer to *readThread()* function. Since this moment program executes two *while* loops, one in main thread and one in the newly created thread. The first one is responsible for processing user input and the second one for displaying incoming serial data. A user by writing *on, off,* or *q* can turn the LED on/off or terminate the application. Unlike on Windows created thread is terminated upon main thread exit so we don't have to terminate it explicitly.

If you are interested in more advanced serial port programming under Linux operating system you are encouraged to look for information about the *select()* function. It is useful for waiting for data from multiple sources, for example if you wait for data from two serial ports and you don't use threads or signals. Additionally if you don't know from which of the serial ports data are coming first using blocking read operation can be tricky. In such a situation *select()* can be a handy solution.

5.4.   .NET, serial port and Linux

As you probably know it is possible to run .NET applications on Linux using Mono. It's an Open Source replacement of Microsoft .NET. In theory you can take .exe file compiled on Windows and run it on Linux. In practice however there are some difficulties.

First of all Mono must be installed on Linux machine. On Ubuntu 11.04 it is available by default but not all assemblies are present. For example you can't run Windows Forms applications. Thus you need to install the *libmono-winforms2.0-cil* package and its dependencies. You can do it using standard command:

```
sudo apt-get install libmono-winforms2.0-cil
```

Now you should be able to run at least simple Windows Forms applications. If you want to compile C# source code, use the *gmcs* command.

As for the serial port programming you can encounter some issues. One of them is enumerating available serial ports using *System.IO.Ports.SerialPort.GetPortNames()*. It works well, in fact too well. The method returns all ttyS* files from /dev/ directory. Not all of them represent physical ports; actually most of them don't resemble physical ports. Virtual serial ports provided by USB adapters are enumerated correctly. But anyway it's no big deal, just the list of available serial ports is longer than it should be. Much more serious issue is lack of support for the event for incoming data. The *SerialPort* class has DataReceived field which is multicast delegate holding list of functions to be called when new data arrive. Usually we provide just one function. On Mono we can't do that, it is not supported and the function won't be called. This is a known issue since 2007 but as of the time of writing this book it hasn't been resolved. The solution is to perform read and write operations sequentially or create separate thread responsible for receiving data. Regardless whether you choose to use threading or not, you must remember that the overloaded *Read()* method which operates on *char* buffer is not implemented – for all incoming bytes null characters are returned. Thus you need to use *byte* data type. Again, this is not big issue but you have to remember this. Otherwise you could spend a lot of time wondering why you get zeros instead of actual data. Mono also doesn't implement *DiscardNull*, *ParityReplace*, and *ReceivedBytesThreshold* fields but they are rarely used so you don't need to care much about them.

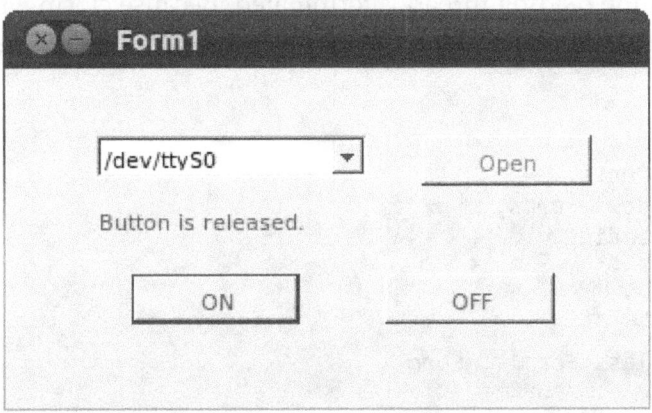

Fig. 5.4.1. Example .NET application running on Linux.

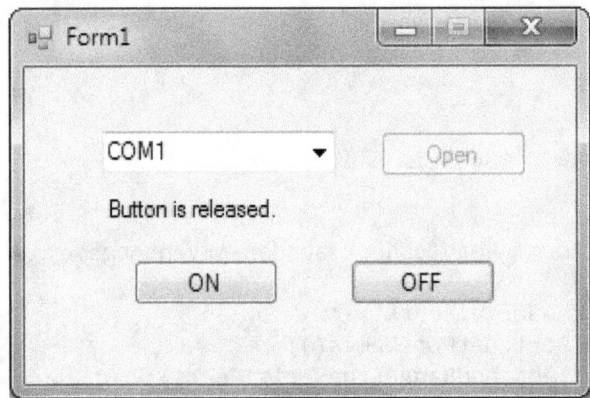

Fig. 5.4.2. The same executable running on Windows.

Let's check out simple program written in a way to be portable and work the same way on Linux as on Windows. It's designed to communicate with a device mentioned in previous chapters, having a button and a LED. It also operates in similar way. Communication is done using just four single characters. A user can select a serial port and open it. Then he/she can turn the LED on and off and observe state of the button. Data are received and interpreted in separate thread. When the application is being closed then the opened port is closed as well and the reading thread is terminated. Because closing a port during read operation causes an I/O exception, it is caught. Here is the code:

```
using System;
using System.IO.Ports;
using System.Windows.Forms;
using System.Threading;

namespace WinLin
{
public partial class Form1 : Form
{

SerialPort ComPort;
string state = "";
        Thread listenerThread;

        public Form1()
        {
            InitializeComponent();
        }

        private void Form1_Load(object sender, EventArgs e)
        {
            string[] theSerialPorts =
System.IO.Ports.SerialPort.GetPortNames();
            cmbPorts.Items.AddRange(theSerialPorts);
            cmbPorts.Text = cmbPorts.Items[0].ToString();
        }
```

```
        private void btnOpen_Click(object sender, EventArgs
e)
        {
            ComPort = new
SerialPort(cmbPorts.SelectedItem.ToString());
            ComPort.BaudRate = 9600;
            ComPort.Parity = Parity.None;
            ComPort.StopBits = StopBits.One;
            ComPort.DataBits = 8;
            ComPort.Handshake = Handshake.None;
            ComPort.Open();
            listenerThread = new Thread(listen);
            listenerThread.Start();
            btnOff.Enabled = true;
            btnOn.Enabled = true;
            btnOpen.Enabled = false;
        }

        private void lblState_DataReceived(object sender,
EventArgs e)
        {
            lblState.Text = "Button is " + state + ".";
        }

        private void Form1_FormClosing(object sender,
FormClosingEventArgs e)
        {
            if (ComPort!=null) ComPort.Close();
            if(listenerThread!=null) listenerThread.Abort();
        }

        private void btnOn_Click(object sender, EventArgs e)
        {
            ComPort.Write("a");
        }

        private void btnOff_Click(object sender, EventArgs e)
        {
            ComPort.Write("b");
        }

        void listen()
        {
```

```
while(true) {
    byte[] buffer = new byte[1];
    int read = 0;
    try
    {
        read = ComPort.Read(buffer, 0, 1);
    }
    catch (System.IO.IOException e)
    {
        break;
    }
    byte val = buffer[0];
    if (val == 'P') state = "pressed"; else state
= "released";
    lblState.Invoke(new
EventHandler(lblState_DataReceived));
    }
    }
    }
}
```

5.5.   *Accessing serial port from console*

Sometimes when you don't have required tools or you want to do something quickly you can take advantage of command line (text shell) of your operating system. On Windows it is cmd.exe. It provides a command for configuring some system setting, e.g. code page of the console or parameters of parallel and serial ports. RS232 ports are seen as special files named COMx where x is a number identifying particular port. There is also a file named AUX which is the same as COM1, similarly PRN points to LPT1. You can check properties of COM1 port by typing:

```
mode com1
```

This will show you information such as speed, parity, or data bits. To change settings add parameters using this pattern:

```
mode com1 9600,n,8,1
```

After you issue this command you can see that settings have been changed. In this example speed of transmission was set to 9600 bits per second, there are no parity bits, there are 8 bits of data, and one stop bit. As it was said, we can treat serial ports as files. Let's say we want to send a file through serial port. This can be simply done with standard copy command:

```
copy file.txt com1
```

This way a file named file.txt was send using COM1 serial port. As you can see COMx files don't have particular location on a disk, you can use them in whatever directory is the current one.

197

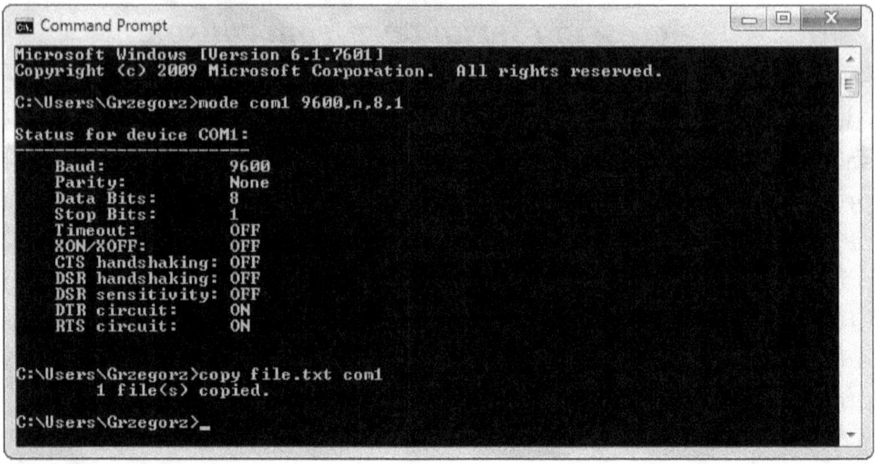

Serial port and microcontrollers

Fig. 5.5.1. Working with serial port using Command Prompt on Windows.

How can we send some data by typing them instead of sending saved file? Again we can use copy command. This time we will need another special file named CON. This file represents a console which means it is responsible for both keyboard and screen. When this file is being read then data are read from the keyboard, when the file is begin written then data are displayed on the screen. So in order to send data typed on a keyboard we can use following command:

```
copy con com1
```

Now you can type some text. When you are finished press Ctrl+Z and the data will be sent.

Reading data from serial port goes in analogous way. If you want incoming data to be written to a file you can enter such a command:

```
copy com1 file.txt
```

To see incoming data on the screen issue this command:

```
copy com1 con
```
or

```
type com1
```

If you start playing with these commands you can run into some problems. One of them is the fact that the transmission is done in ASCII (text) mode. This means that data in a file are interpreted before they are sent. For example if you try to send a file which contains a byte of value 26 (1Ah) then following bytes won't be sent. Such behavior is caused by interpreting this byte as end of file marker. If we want to send a file as it is we have to add /b switch:

```
copy /b file.bin com1
```

Another problem is also connected with indication of an end of file. If you receive data using type or copy commands then they constantly wait for data to come. We can break it by pressing Ctrl+C but this aborts whole operation and the copy command won't save received data to a file. In order to stop reading incoming bytes we have to supply end of file byte (value 26). It just has to be sent by the other device. This means that the copy command isn't suitable for receiving binary data.

You may also notice one more problem: received data are buffered. This means that they are displayed or written to a file only after buffer overflow. Size of this buffer is 256 bytes. Windows doesn't provide any easy way to disable the buffering. Fortunately it turns out that it is disabled by opening serial port using another application, like PuTTY or simple C# code presented earlier in this book. Buffering can be enabled again by restarting Windows or disabling and enabling serial port in device manager.

Let's summarize limitations of Windows command prompt (cmd.exe) with respect to transmitting files:

- to send binary files use /b switch

- to receive files make sure buffering is disabled

Serial port and microcontrollers

- if you save received data to a file, one additional byte of value 26 (1Ah) marking end of file must be sent

With Windows command prompt, and many other shells as well, we can not only transmit files or some short texts but also redirect output from executed commands. As you probably know output of a command can be redirected to a file. Then the standard output from the command goes not to screen but to disk. We do it using the ">" operator, e.g.:

```
dir > directory_listing.txt
```

This gives us directory listing saved to a file, not shown on a screen. As COM ports are treated as files we can simply type:

```
dir > com1
```

This way result of *dir* goes through serial port and can be received by another device, e.g. a PC. Output of *dir* received with PuTTY through RS-232 was shown on figure 5.2 (beginning of chapter 5).

cmd.exe is not the only text shell offered by Microsoft for Windows. There is also a new, object-oriented one called PowerShell. It's much more powerful and versatile than standard *cmd.exe*. It's available for Windows XP and newer versions of Windows, in Windows 7 it is available out of the box. In PowerShell you can use .NET classes so working with serial port can very similar to programming in C#. This book is not about programming in PowerShell so we won't be going into details. On the other hand this shell is very handy and you might want to use it for writing useful scripts making working with Windows easier. So it's good to know how to use it for transmitting data over serial port.

As it was said, serial port can be accessed using .NET classes so the code will be very similar. For enumerating available serial ports we can type:

```
[System.IO.Ports.SerialPort]::getportnames()
```

A port object can be created this way (in one line):

```
$port = new-object System.IO.Ports.SerialPort
COM1,9600,None,8,One
```

This way we have initialized an object tied to first serial port with speed set to 9600 bps, no parity check, 8 bytes of data and one stop bit. To open the port we just issue following command:

```
$port.open()
```

Now we can send and receive data. For example to send line of text we can use the WriteLine() method:

```
$port.WriteLine("Hello World!")
```

Similarly to receive a line of text we can type:

```
$port.ReadLine()
```

Remember that by default the *ReadLine()* method waits for CR and LF characters. For example when you hit Enter in PuTTY then only CR is sent so you need to modify value of *NewLine* field of the serial port object.

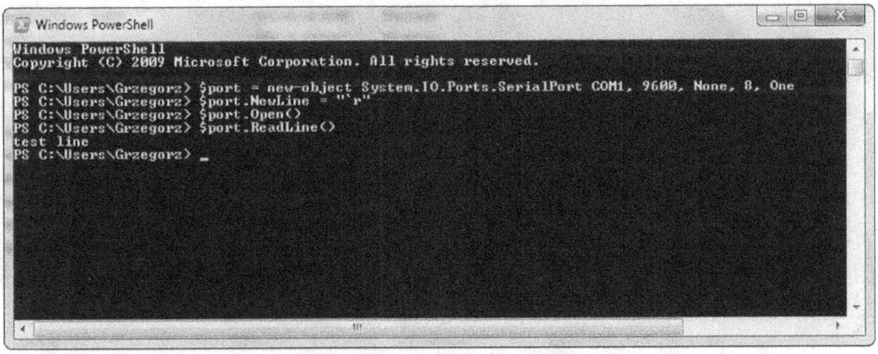

Fig. 5.5.2. Receiving line of text using PowerShell.

Serial port and microcontrollers

If you want to receive data asynchronously you have to register a handler using Register-ObjectEvent cmdlet. Source code which processes data is provided as Action parameter (one line):

```
Register-ObjectEvent -inputobject $port -eventname
"DataReceived" -Action { write-host $port.readexisting() }
```

Now you know how to use the shells provided with Windows to perform basic operations with serial port. Let's take a look how this looks like in Linux. Similarly there are device files but they have specified location: the */dev* directory. The first serial port is available there as *ttyS0* file, the second one as *ttyS1* and so on. If you have USB to serial adapter then the first virtual serial port is called *ttyUSB0*, if there is a second one it is *ttyUSB1* and so on. Parameters of transmission can be configured with *stty* command (one line):

```
stty -F /dev/ttyS0 ispeed 9600 ospeed 9600 -ignpar cs8 -
cstopb -ixon
```

This command operates on first serial port (*ttyS0*) and sets input and output speed to 9600 bits per second. Parity checking is disabled, number ot data bits is set to 8, number of stop bits is set to one and XON/XOFF flow control is disabled. Detailed description of *stty* command is available in manual:

```
man stty
```

When working with *ttySX* you have to remember about default settings. First of all echo is enabled so everything coming to the port is sent back. This sometimes can be helpful when we work work remote terminal which doesn't provide local echo and we can't see what we type. In other cases it may be undesired. Another thing is text mode which changes line endings from <LF> to <CR><LF>. It can be disabled by switching to raw mode.

```
stty -F /dev/ttyS0 raw
```

This mode also disables buffering for incoming data. When raw is not enabled, for example for sane mode, data are passed only after end of line marker is received. You can observe it after connecting with PuTTY or similar program, your letters won't appear on Linux console until you press Enter. *raw* and *sane* aren't actually single parameters but they group many parameters together. It may happen that you will have to manipulate individual parameters, e.g. *echo*. Remember that adding a dash "-" disables a parameter, omitting a dash before name of parameter enables it.

Sending and receiving data can be performer in analogous way as in Windows. To send a file we can use cp command:

```
cp file.txt /dev/ttyS0
```

And to receive a file:

```
cp /dev/ttyS0 file.txt
```

To stop receiving data press Ctrl+C. To display incoming data you can use *cat* or *cp* commands:

```
cat /dev/ttyS0
```

```
cp /dev/ttyS0 /dev/tty
```
What is */dev/tty*? It points to current console. If you want to specify the console explicitly, you can get current console under bash with following command:

```
echo $(tty)
```

If you want to send some data from keyboard you can again use cp command:

```
cp /dev/tty /dev/ttyS0
```

It must be noted that whole lines will be sent, not individual characters. To finish sending text and return to prompt press Ctrl+C.

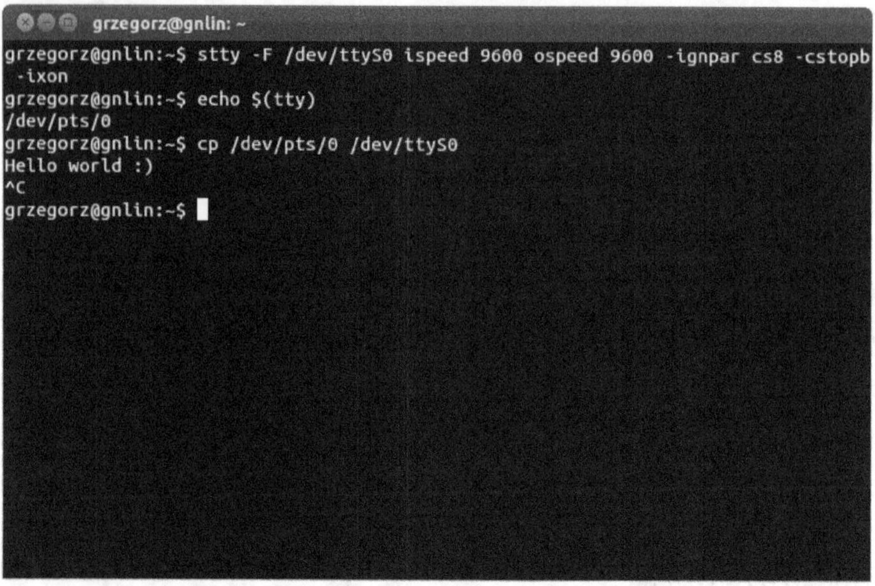

Serial port and microcontrollers

```
grzegorz@gnlin: ~
grzegorz@gnlin:~$ stty -F /dev/ttyS0 ispeed 9600 ospeed 9600 -ignpar cs8 -cstopb
 -ixon
grzegorz@gnlin:~$ echo $(tty)
/dev/pts/0
grzegorz@gnlin:~$ cp /dev/pts/0 /dev/ttyS0
Hello world :)
^C
grzegorz@gnlin:~$ 
```

Fig. 5.5.3. Accessing serial port with Linux console.

Sometimes it's necessary to send some special characters, for example new line or end of file. We can use echo and escape codes for that. To send <CR> and <LF> we use the same codes as in C or C#: "\r" for carriage return and "\n" for line feed. Thus to send <CR><LF> one can use following command using bash:

```
echo $'\r\n' > /dev/ttyS
```

To send end of file marker you can type:

```
echo $'\cZ' > /dev/ttyS0
```

For other escape sequences please refer to bash documentation.

5.6.   *Using serial port in web applications*

Nowadays web applications are very common and we use them more and more frequently. If we have same device which we want to control or get data from, it may be a good idea to write a web application for easy remote access.

Unfortunately this task is not so simple. This is because of the fact that webpages were invented for presentation of linked documents, not as an application platform. The model was simple: a browser sends a request and a server sends back a webpage. This way data could be sent to the server any time but the server can send data to the client only after a request. As Web started to became serious application platform the situation got better but is still an issue. The problem of sending data to client asynchronously is circumvented in several ways:

- with AJAX request (XMLHttpRequest) which perform periodic polling of a server

- persistent connection, server keeps a connection open and sends chunks of data

- HTML5 WebSocket API

- variations of first two mechanisms

Generally the web application model where server pushes data back to a browser without an explicit request, regardless of implementation, is called Comet. If you want your web application to present real time data you have to examine existing Comet approaches to the problem and choose the most suitable for your needs.

The other problem is lifecycle of a web application. The server-side part of an application is executed for a short time, usually under a second, when user waits for a webpage to be generated and send to him/her. Moreover servers kill scripts which take too long to complete, usually the timeout is 60 seconds. On the other hand the client-side part can be executed as long as a user has the webpage opened in a browser. We are interested in server-side code because it is going to be responsible for RS-232 communication. As it must be executed quickly we can't place there a call to a read function which would block execution for several minutes or more. In order to solve the problem of fetching data from an RS-232 device we have two options:

- use a device which returns all interesting data upon request so read operations can complete in a very short time

- create additional application working as a proxy between a serial device and a web application

Which approach to use depends on the type of device you want to support, or more precisely, on its communication protocol. In chapters about programming desktop application there was mentioned a device which sends state of a button (pressed/released) in the moment of change of that state. For such a case the second solution is needed, where special application is responsible for receiving data and making them accessible for a web application. Anyway look at what the platform you use can offer you. In ASP.NET serial port is handled the same way as presented in this book in the chapter about C#. For PHP there as several serial port classes available on the Web for both Windows and Linux.

OK, so let's see how we can create a webpage which can communicate with serial port through a dedicated application. Presented solution shows state of a button (pressed or not pressed) and allows to turn on/off a LED. It's just an example of a task which may be realized in many different ways. Here we are going to use a Linux server witch Apache, MySQL, and an application reading and writing to a serial port. A MySQL database stores data received by the application and data designed to be sent. Apache serves PHP website for communication with remote user. This website can talk to the database to interchange information with the application responsible for serial port.

The source code of the application which handles serial port communication is shown below. It's a main component of the whole solution and it shows how it works.

```
#include <iostream>
#include <my_global.h>
#include <mysql.h>
#include <string.h>
#include <sys/types.h>
#include <sys/stat.h>
#include <fcntl.h>
#include <termios.h>
#include <sys/signal.h>
#include <unistd.h>

#define SERIAL_PORT "/dev/ttyS0"
#define BAUDRATE B9600
#define PRESSED_QUERY "UPDATE btn_state SET state=true"
#define NOTPRESSED_QUERY "UPDATE btn_state SET state=false"

void setupdb(MYSQL * conn);
int initSerial();
void signal_handler_IO (int status);
int getNewDataDB(MYSQL * conn, int & state);

int newData = 0;

using namespace std;

int main(int argc, char * argv[]) {
    MYSQL * conn = mysql_init(NULL);
    if(!mysql_real_connect(conn, "localhost", "grzegorz",
"password1", "serial", 0, NULL, 0)){
        printf("Error %u: %s\n", mysql_errno(conn),
mysql_error(conn));
        exit(1);
    }
```

```
    if (argc==2) if (strcmp(argv[1], "setup")==0)
setupdb(conn);
    int fd = initSerial();
    while(1) {
        if(newData){
            newData = 0;
            char buf[1];
            read(fd, buf, 1);
            cout << buf[0] << endl;
            if (mysql_query(conn, buf[0]=='P' ? PRESSED_QUERY :
NOTPRESSED_QUERY)) {
                printf("Error %u: %s\n", mysql_errno(conn),
mysql_error(conn));
                exit(1);
            }
        }
        int state;
        if(getNewDataDB(conn, state)){
            char buf[] = {'a'};
            if(!state) buf[0] = 'b';
            write(fd, buf, 1);
            newData = 0;
        }
        usleep(100000);
    }
    mysql_close(conn);
    return 0;
}

void setupdb(MYSQL * conn) {
    mysql_query(conn, "drop table btn_state");
    mysql_query(conn, "drop table led_state");
    if (mysql_query(conn, "create table btn_state (state
BOOL)")) {
        printf("Error %u: %s\n", mysql_errno(conn),
mysql_error(conn));
        exit(1);
    }
    if (mysql_query(conn, "insert into  btn_state (state)
values (false)")) {
        printf("Error %u: %s\n", mysql_errno(conn),
mysql_error(conn));
        exit(1);
```

```
    }
    if (mysql_query(conn, "create table led_state (state
BOOL, new BOOL)")) {
        printf("Error %u: %s\n", mysql_errno(conn),
mysql_error(conn));
        exit(1);
    }
    if (mysql_query(conn, "insert into  led_state (state,
new) values (false, false)")) {

    }
}

int initSerial() {
    int fd = open(SERIAL_PORT, O_RDWR | O_NOCTTY);
    struct sigaction saio;
    saio.sa_handler = signal_handler_IO;
    sigemptyset(&saio.sa_mask);
    saio.sa_flags = 0;
    sigaction(SIGIO, &saio, NULL);
    fcntl(fd, F_SETOWN, getpid());
    fcntl(fd, F_SETFL, FASYNC);
    struct termios newtio;
    bzero(&newtio, sizeof(newtio));
    newtio.c_cflag = BAUDRATE | CS8 | CLOCAL | CREAD;
    newtio.c_iflag = 0;
    newtio.c_oflag = 0;
    newtio.c_lflag = 0;
    newtio.c_cc[VTIME] = 0;
    newtio.c_cc[VMIN] = 1;
    tcflush(fd, TCIFLUSH);
    tcsetattr(fd, TCSANOW, &newtio);
    return fd;
}

void signal_handler_IO(int status) {
    newData = 1;
}

int getNewDataDB(MYSQL * conn, int & state) {
    mysql_query(conn, "SELECT state, new FROM led_state");
    MYSQL_RES * result = mysql_store_result(conn);
    MYSQL_ROW row = mysql_fetch_row(result);
```

```
    int newState = atoi(row[1]);
    mysql_free_result(result);
    if(newState) {
       state = atoi(row[0]);
       if (mysql_query(conn, "UPDATE led_state SET
new=false")) {
          printf("Error %u: %s\n", mysql_errno(conn),
mysql_error(conn));
          exit(1);
       }
       if(state) cout << "LED ON" << endl; else cout << "LED
OFF" << endl;
    }
    return newState;
}
```

As you can see at the beginning of the *main()* function, the program connects to a local MySQL instance using username and password and selects *serial* database. The application uses two tables in the databse: *btn_state* and *led_state*. They must be present in the database. If they are not present we can create them by launching the application with *setup* argument. Then the *setupdb()* function creates them and fills with default values. Such things as user account (here it's *grzegorz*) and database (here *serial*) have to be created using appropriate MySQL administration tool, e.g. MySQL Administrator.

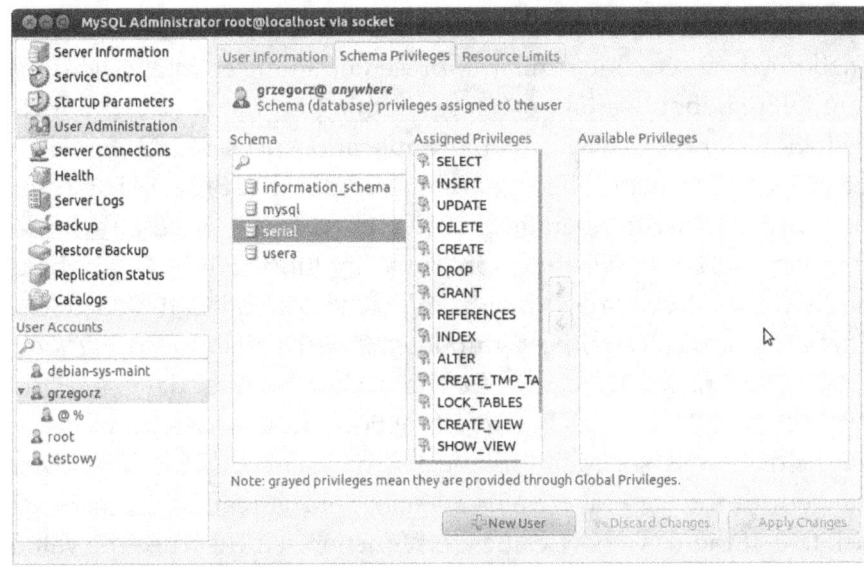

Fig. 5.6.1. MySQL Administrator, here used to give user grzegorz permissions to database serial

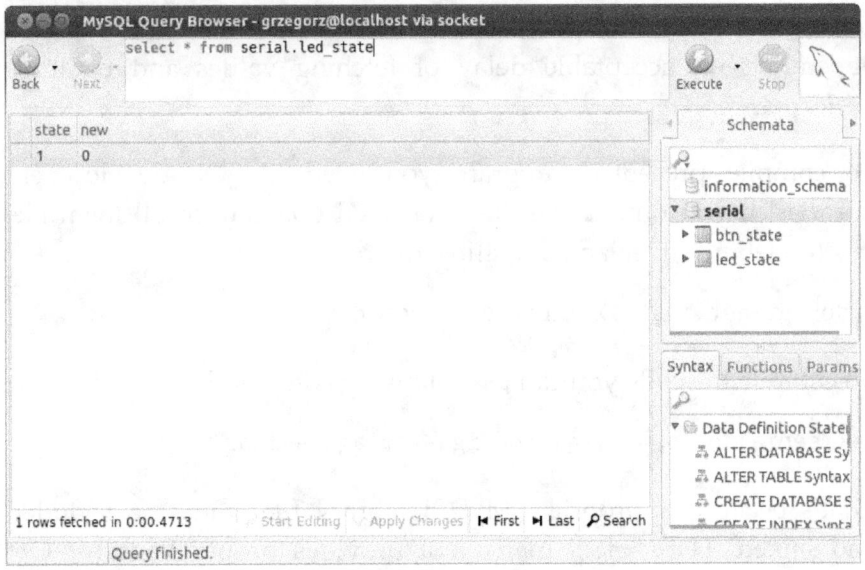

Fig. 5.6.2. MySQL Query Browser allows viewing and manipulating data in databases, invaluable during testing

If you look at the rest of the *main()* function and at the *initSerial()* function you'll see code similar to presented in the chapter where we were talking about serial port programming on Linux. Serial port is used in a standard way and the application uses SIGIO signal to detect incoming data. The important thing is that SIGIO is sent for both read and write operations so we need to set *newData* to zero after write. The *main()* function runs an infinite loop. This loop checks constantly for new data from serial port and in the database. If data come from serial port it updates the *state* field in the *btn_state* table accordingly. To check for new data in the database, the application calls *getNewDataDB()* function. First it checks value of *new* field. If it is set to 1 this means that state of LED should be changed. In this case also value of *state* field is read and *new* is set back to 0. If there is a new value then function returns true. The value itself is returned by the *state* parameter which is passed by a reference. At each iteration of the main loop execution of the program is halted for 100 000 microseconds (0.1 seconds) which gives 10 database queries per second. This value may be changed depending on acceptable delay of fetching values and database performance.

To compile presented program you need MySQL development libraries. If you don't have them installed you can install them like this (on Ubuntu/Debian-compatible distros):

```
sudo apt-get install libmysqlclient-dev
```

To compile the code you can use following command:

```
c++ serialdp.cpp `mysql_config --cflags --libs`
```

That's because appropriate MySQL library files must be included and linked. Here *mysql_config* tool provides proper flags for C++ compiler. It's important to have source file name before flags. Otherwise linker may perform too aggressive optimization and not link MySQL libraries, which gives errors like this:

```
serialdp.cpp:(.text+0x12): undefined reference to
`mysql_init'
```

Having the database configured and application compiled you can test it. It's good to do it before you start creating a webpage so you can detect problems early and not get confused by problems caused by the webpage. First try to run the program with *setup* parameter and check if both tables have been created successfully. If yes then run the program without any parameters and check if it reacts correctly for incoming 'P' and 'N' characters from the serial port. It should update the *btn_state* table accordingly. If everything is OK then check if program notices changes in *led_state* table properly. When the application works correctly you can move to the web part.

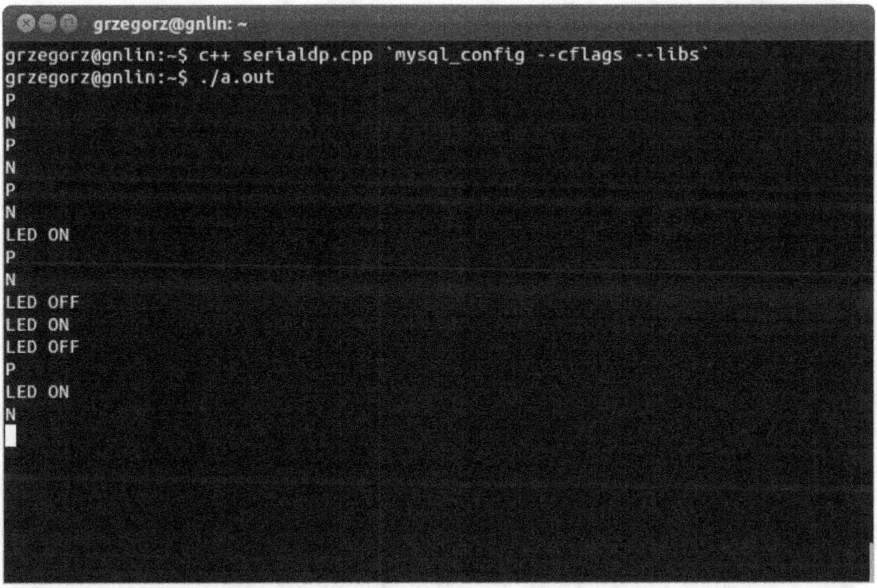

```
grzegorz@gnlin: ~
grzegorz@gnlin:~$ c++ serialdp.cpp `mysql_config --cflags --libs`
grzegorz@gnlin:~$ ./a.out
P
N
P
N
P
N
LED ON
P
N
LED OFF
LED ON
LED OFF
P
LED ON
N
```

Fig. 5.6.3. Running application responsible for communication with serial port

As it was said, in this example we are using Apache and PHP. Be sure you have them installed and running. Don't forget about MySQL libraries for PHP:

```
sudo apt-get install php5-mysql
```

The web application is very simple. It has two functions: shows current button state and allows turning LED on and off. Let's look at the code:

```php
<?
    if (mysql_connect(NULL, "grzegorz", "password1") and
mysql_select_db("serial")) {
        $result = mysql_query("SELECT state FROM led_state");
    } else echo "Can't connect to database!";
    while($rec = mysql_fetch_array($result)) {
        if($rec['state']) $checked = "checked=true";
    }
?>

<html>
<head>

<script language="javascript">

var xmlHttp = new XMLHttpRequest();

function getButtonState(){
    xmlHttp.open("POST", "get_button.php", true);
    xmlHttp.onreadystatechange = handleRequestStateChange;
    xmlHttp.send(null);
}

function handleRequestStateChange(){
    // continue if the process is completed
    if (xmlHttp.readyState == 4) {
      // continue only if HTTP status is "OK"
      if (xmlHttp.status == 200) {
          var response = xmlHttp.responseText;
            if(response!=0)
              document.getElementById("button").innerHTML =
"pressed";
            else
              document.getElementById("button").innerHTML =
"not pressed";
```

```
        }
      }
}

function checkboxClick(){
    var xmlHttpSet = new XMLHttpRequest();
    xmlHttpSet.open("POST", "set_led.php", true);
    xmlHttpSet.setRequestHeader("Content-type",
"application/x-www-form-urlencoded");
    var ledstate = document.getElementById("led").checked;
    xmlHttpSet.send("led_on="+ledstate);
}

setInterval("getButtonState()", 500);

</script>

</head>

<body>

button state: <span id="button"></span>
<br/>
<form>
LED: <input type="checkbox" id="led"
onclick="checkboxClick()" <? echo $checked ?> />
</form>

</body>
</html>
```

To show state of the button we use a *span* element. Its text is updated using AJAX requests. With *setInterval()* function we call *getButtonState()* function every 500 milliseconds. This function creates an AJAX request which downloads result of *get_button.php* script. This script is very simple, it just retrieves button state from the database. Its code is shown here:

```
<?
    if (mysql_connect(NULL, "grzegorz", "password1") and
mysql_select_db("serial")) {
        $result = mysql_query("SELECT * FROM btn_state");
    } else echo "Can't connect to database!";
    while($rec = mysql_fetch_array($result)) {
        echo $rec['state'];
    }
?>
```

It's a typical PHP code for getting data from a MySQL database. First a connection is made using appropriate login and password, and then the database is selected. When the connection has been established a SELECT query is issued to obtain a row from the *btn_state* table. The *state* field is returned and passed back to the AJAX request. When it's finished then the handleRequestStateChange() callback function is executed. It updates the *span* element using the value returned by *get_button.php* with AJAX.

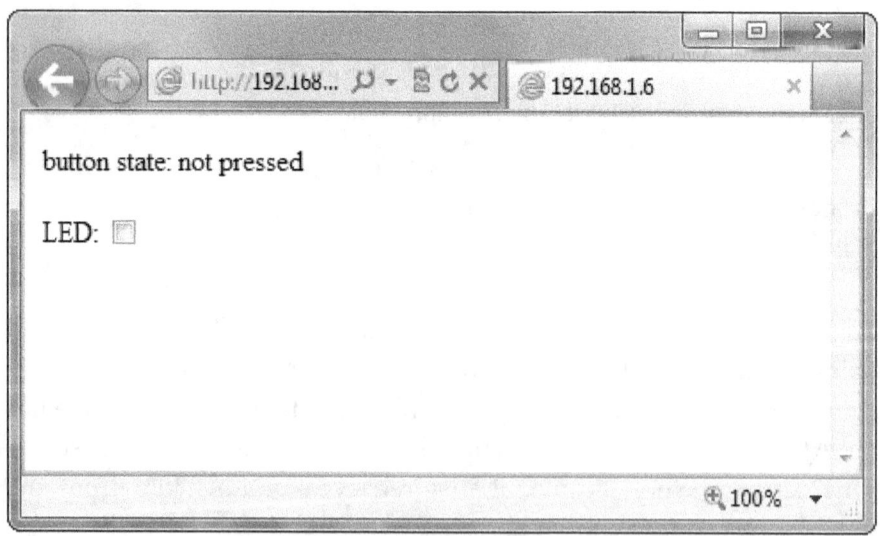

Fig. 5.6.4. View of the web application

A short explanation why the used AJAX request is a POST type, not GET type. In theory it should be GET because we are not posting any data. We use POST for compatibility with Internet Explorer which uses very aggressive caching for AJAX requests. With this caching IE doesn't make connections to a server for subsequent requests but reuses the value obtained previously and stored in browser's cache. There are several workarounds for this problem, e.g. adding random parameters to URL. One of the easiest is just to use POST instead of GET. It must be also noted that presented code is not compatible with IE6 and earlier because of the way the AJAX object is being created. If you happen to be required to use such old version of IE look for information how to instantiate AJAX object as ActiveX.

The other function realized by presented web application is turning LED on and off. It is realized in a very similar way. We have a checkbox which launches *checkboxClick()* function which creates an AJAX request to *set_led.php*. State of the checkbox is passed as a *led_on* parameter. The code of *set_led.php* is simple and analogous to *get_button.php*:

```
<?
    $state = $_POST['led_on'];
    $query = "UPDATE led_state SET state=".$state.",
new=true";
    if (mysql_connect(NULL, "grzegorz", "password1") and
mysql_select_db("serial")) {
        echo $query;
        $result = mysql_query($query) or die("Error in
query!");
        echo $result;
    } else echo "Can't connect to database!";
?>
```

First the code retrieves passed *led_on* parameter and prepares appropriate SQL statement. Then a connection is made and the *led_state* table is updated.

217

It's pretty much everything. The last thing to mention is that when we open the webpage then it would be nice to have the LED checkbox checked or not according to current state of the LED. Therefore on the beginning of the page's code we fetch the state from the database. If the LED is turned on then we set the *$checked* variable to store *checked* attribute for the checkbox. This variable is later inserted into *input* tag.

As it was sad at the beginning presented approach is just a simple example. Simple and having its downsides, for example frequent polling of both database and webserver. However it could be a good start to create own, more advanced solutions, using different architecture, technologies, and tools.

6. Examples

6.1.   First communication

When we know how to write simple programs for AVRs and ARMs it's time to make first communication between our microcontroller and a computer through RS-232. Our first task will be turning a LED on and off using an application running on a PC. This application will be sending commands to our microcontroller and it will respond by changing state of the diode. It's a simple example of unidirectional communication.

6.1.1. Hardware configuration

First, you need to make sure that basic connections are made: you microcontroller is connected to power (pins VCC, GNC, AVCC and AGND for AVR, and VDD_1/2/3, VSS_1/2/3, VSSA, and VDDA for STM32) and to programming interface (ISP or JTAG). In case of STM32 you have also to take care about BOOT0 and BOOT1/PB2 pins. If you wish to use a crystal resonator as clocking signal you have to connect it with accompanying capacitors as well. All those connections are described in the chapters devoted to hardware design. If you use a development board just connect LED and check power and JTAG/ISP connections.

The LED can be connected to any free pin of one of available I/O ports. In our example we use pin 0 of port A (for both AVR and STM32). It is pin 40 of ATmega32 in DIP package and pin 10 of STM32F103CBT6 in LQFP package. The LED has to be connected in series with a resistor, between pin of the port and ground. The resistor limits current flowing through the LED diode to a value acceptable by this element. Modern LEDs need just a few milliamperes so a 1 kOhm resistor is a good choice. You can use lower value for brighter light higher for dimmer light and smaller power consumption. Make sure however, that you don't exceed maximal current tolerable by port of a microcontroller or LED.

Actual values you can find in datasheets of those elements. As the LED is connected in series with the resistor it doesn't matter which element is on the microcontroller side or ground side. Nonetheless you need to take care about polarity, the cathode needs to be on ground side and anode on microcontroller side. The LED can be connected also in other way: between pin and supply voltage VCC. This time cathode faces a port pin and anode faces VCC. Of course there must be a serial resistor between the microcontroller and LED or between LED and VCC, just like in first case. The difference between those two modes is driving the pin by the software. When the diode is connected to GND, you need to set the port pin to 1 in order to turn on the light emitting diode. When the diode is connected to VCC you have to do the opposite: set port pin to 0 to turn the LED on or to 1 to turn it off.

The RS-232 voltage converters were described in the chapter covering RS-232 standard. Probably you are going to use MAX3232 or one of its clones as it is the most popular voltage converter for RS-232. As it has two sets of converters, you can choose any one you want. This means that you can connect RXD pin of a microcontroller to either R1 OUT (pin 12) or R2 OUT (pin 9) of MAX3232. Of course if you use R1 OUT, you have to connect R1 IN to the RS-232 line and if you use R2 OUT then R1 IN should be connected to the RS-232 line. In our first example we are not sending any data to a PC so we don't have to take care about TXD pin but it's not bad to have it connected for further experiments. The situation is analogous: TXD can be connected to either T1 IN (pin 11) or T2 IN (pin 10). If you use T1 IN then T1 OUT must be used on RS-232 line side, T2 OUT in other case. In ATmega32 in DIP package the RXD pin is pin 14, which can also work as pin 0 of port D. The TXD pin is pin 15, which can work as pin 1 of port D. In case of STM32F103CBT6 in LQFP package the RXD is pin 31 (pin 10 of port A) and TXD is pin 30 (pin 9 of port A). For detailed information about MAX3232, including connecting required capacitors, please refer to one of previous chapters.

If you are going to use the ZL15AVR development board or similar, most of needed connections are already present. You just need to make three connections:

- connect PA0 to one of LEDs

- connect PD0 to RxD

- connect PD1 to TxD

For ZL30ARM do as follows:

- connect PA0 to one of LEDs

- connect PA10 to RxD

- connect PA9 to TxD

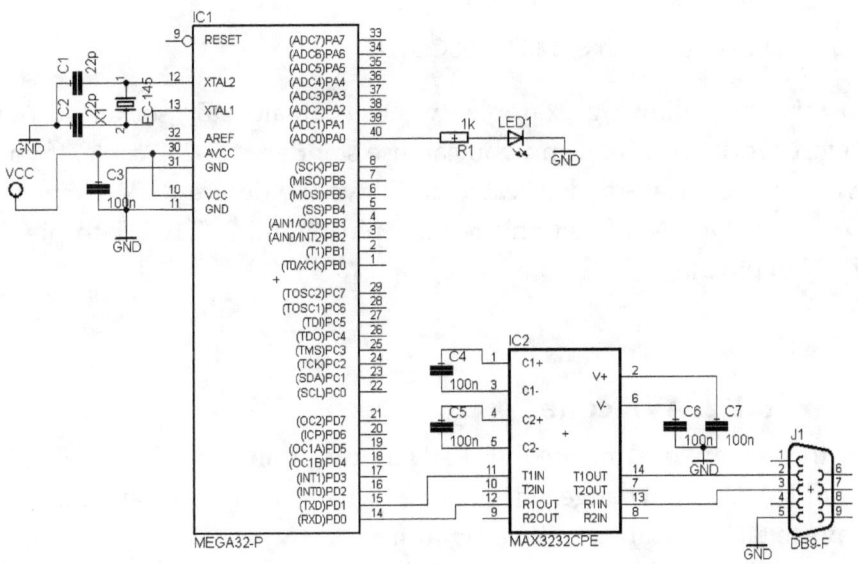

Fig. 6.1.1.1. Circuit with ATmega32. JTAG/ISP connections omitted.

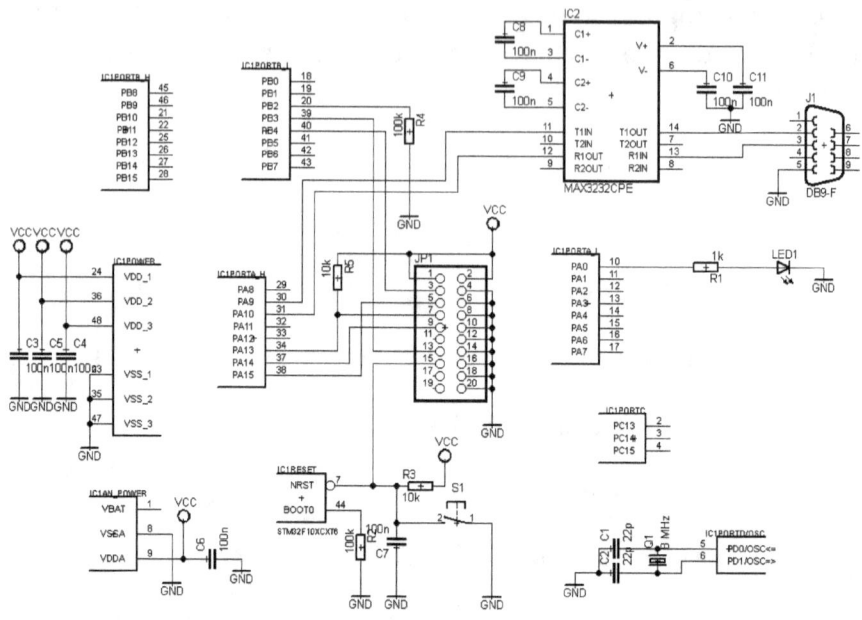

Fig. 6.1.1.2. Circuit with STM32F103CBT6.

In this and following examples we use female DB9 socket so our device works as a DCE and you can use standard RS-232 cable. If you want to use a null-modem cable and have the device to be seen as a DTE, take DB9-M socket and replace pins 2 and 3 (T1OUT to pin 3, R1IN to pin 2).

6.1.2. AVR code

In this example a microcontroller has two things to do: read data from UART and change state of one of its pins. This means that we have to take care about following things:

- initialize UART: turn UART on, configure it for receiving data, set speed, parity and stop bits, enable UART interrupt

- create UART interrupt handler: write function called when data comes, which would analyze these data and drive LED accordingly

- initialize I/O port: one of pins we choose must be configured as an output

Let's look at sample implementation of above steps:

```
#include <avr/io.h>
#include <avr/interrupt.h>

void init_UART (void)
{
    UBRRL = 103; //9600 bps @ 16 MHz crystal
    UCSRB = _BV(RXEN)|_BV(RXCIE); //enable receiving data,
enable interrupts
    UCSRC = _BV(URSEL)|_BV(UCSZ1)|_BV(UCSZ0); //n,8,1
}

ISR(USART_RXC_vect) {
    uint8_t data = UDR;
    if (data=='a') {
        PORTA |= _BV(PINA0);
    }
    if (data=='b') {
        PORTA &= ~_BV(PINA0);
    }
}

int main() {
    init_UART();
    sei();
    DDRA = _BV(DDA0);
    for(;;);
}
```

The main() function starts with a call to a function which realizes initialization of UART. First, we are setting speed of transmission, this setting is held in UBRRL register. The value written there depends on desired speed and clock speed and can be calculated using an appropriate equation. You can find that equation in the datasheet of your microcontroller. In practice however it's enough to just look at a table of *UBRR* values which can be also found in the datasheet. From that table we can learn that for speed of 9600 bits per second and 16 MHz clock the value should be 103. For other speed and clock the value will be different. Just remember to use the same speed in your code for the microcontroller and in an application on a PC. Notice that for fast clock and slow speed the value won't fit in *UBRRL* and you will have to use *UBRRH*. For example for 16 MHz clock and 2400 bps speed the value of UBRR is 416 which gives 160 in *UBRRL* and 1 in *UBRRH*. That's because *UBRRH* holds more significant part of *UBRR* and 416 = 256 x 1 + 160. In our examples the *UBRR* value is 103 which fits entirely in *UBRRL* and we don't need to set *UBRRH* value.

In the next line we are setting two bits in the *UCSRB* register. The *RXEN* bit enables UART receiver, pin 0 of port D is reconfigured to work as serial input. The *RXCIE* bit enables interrupts generated when data are received. With *UCSRC* register we are configuring parameters of transmission. By setting *UCSZ1* and *UCSZ0* bits to 1 and *UCSZ2* to 0 we are selecting 8-bit transmission mode. In ATmega32 there can be from 5 to 9 transmitted bits in a frame. For appropriate combinations of *UCSZ0*, *UCSZ1* and *UCSZ2* for these modes please refer to microcontroller's datasheet. By not setting the *USBS* bit we are selecting one stop bit transmission. In the same way, by not setting *UPM1* and *UPM0*, we are turning of parity check. We are also setting *URSEL* bit. It must be set every time *UCSRC* is to be written. This comes from the fact that the *UBRRH* register shares the same I/O location as the *UCSRC*. If you leave *URSEL* set to 0 then *UBRRH* will be updated instead of *UCSRC*.

When USART is configured, the *main()* function enables interrupts globally by calling the *sei()* function. As you can see interrupts have to be enabled in two places: globally and specifically for USART. The next thing to be done is configuration of pin 0 of port A as output. It is done in standard way by writing to the *DDRA* register. The next thing the program does is entering an infinite loop. From this moment the microcontroller does actually nothing, the only thing which can happen is an interrupt coming from USART. Then the handler is executed and the microcontroller returns to jumping inside the infinite loop.

The interrupt handler for incoming data is quite simple. It is called when receiving a data frame (a byte in practice) is finished. At this moment we can read the received value from the *UDR* register. It is then copied to a local variable. It is not done for a convenience or something however. *UDR* is connected to a circular shift register and every time it is read the shift register is shifted and we read another value. The shift register works as circular buffer which stores received data. Thus we don't have to read incoming bytes one by one, each time they are received. A program can wait for a few bytes to be received and then read them all from the buffer. In our example however this approach is not used, we read single value and analyze it with simple if clauses. If the received value is an 'a' character then the pin 0 of port A is set to 1 which turns on connected LED diode. Similarly, if there is 'b' received, LED is turned off by setting pin 0 to 0.

As you can see, there is a transmission protocol defined. Every time an 'a' character is received the LED is turned on, every time a 'b' character is received the LED is turned off. All other values are ignored. The 'a' character is of course the same as a numeric value 97 and 'b' is 98 so the code can be written as follows:

```
if (data==97) {
      PORTA |= _BV(PINA0);
}
if (data==98) {
      PORTA &= ~_BV(PINA0);
}
```

225

If you like, you can use hexadecimal notation, then you will get 0x61 and 0x62. It's up to you. The notation with characters was chosen because for test purposes it's easy to use serial terminal application like Hyper Terminal or PuTTY instead of writing dedicated program. And in such an application you work just by typing on keyboard and sending letters and digits to the RS-232 port of your computer. So if you work with characters it is natural to use character notation instead of numbers.

You can come up with own protocol for this example, e.g. '0' and '1' instead of 'a' and 'b'. Just remember that character '0' doesn't have ASCII value 0 but 48 (30h) and '1' is 49 (31h). In fact you can use any values which fit in one byte but you have to have a way of sending them. For sending values which are not expressed by printable characters you need to write own application instead of using terminal applications (e.g. PuTTY).

Another thing you can change is how the microcontroller responds to received bytes. For example it can toggle state of its pin every time a byte is received.

6.1.3. STM32 code

The serial communication on STM32 platform is similar to the one realized on AVRs. The difference is that most of the code is for initialization of peripherals. This is due to the fact that peripherals on STM32 platform are more sophisticated and, more importantly, because of the design of the STM32 Standard Peripheral Library. In this library filling large configuration structures is not uncommon. As it was said in the chapter devoted to that library, you are free to just write to appropriate registers. Because of readability of the library samples presented here are using it. OK, let's take a look at the code.

```c
#include "stm32f10x.h"

int main(void)
{
  RCC_APB2PeriphClockCmd(RCC_APB2Periph_GPIOA |
RCC_APB2Periph_USART1, ENABLE);

  GPIO_InitTypeDef GPIO_InitStructure;
  GPIO_InitStructure.GPIO_Pin = GPIO_Pin_10;
  GPIO_InitStructure.GPIO_Mode = GPIO_Mode_IN_FLOATING;
  GPIO_InitStructure.GPIO_Speed = GPIO_Speed_2MHz;
  GPIO_Init(GPIOA, &GPIO_InitStructure);

  GPIO_InitStructure.GPIO_Pin = GPIO_Pin_0;
  GPIO_InitStructure.GPIO_Mode = GPIO_Mode_Out_PP;
  GPIO_Init(GPIOA, &GPIO_InitStructure);

  USART_InitTypeDef USART_InitStructure;
  USART_InitStructure.USART_BaudRate = 9600;
  USART_InitStructure.USART_WordLength = USART_WordLength_8b;
  USART_InitStructure.USART_StopBits = USART_StopBits_1;
  USART_InitStructure.USART_Parity = USART_Parity_No;
  USART_InitStructure.USART_HardwareFlowControl =
USART_HardwareFlowControl_None;
  USART_InitStructure.USART_Mode = USART_Mode_Rx;
  USART_Init(USART1, &USART_InitStructure);
  USART_ITConfig(USART1, USART_IT_RXNE, ENABLE);
  USART_Cmd(USART1, ENABLE);

  NVIC_PriorityGroupConfig(NVIC_PriorityGroup_0);
  NVIC_InitTypeDef NVIC_InitStructure;
  NVIC_InitStructure.NVIC_IRQChannel = USART1_IRQn;
  NVIC_InitStructure.NVIC_IRQChannelSubPriority = 0;
  NVIC_InitStructure.NVIC_IRQChannelCmd = ENABLE;
  NVIC_Init(&NVIC_InitStructure);

  while(1);

}

void USART1_IRQHandler(void)
{
  if (USART_GetITStatus(USART1, USART_IT_RXNE) != RESET)
```

```
{
  char data = USART_ReceiveData(USART1);
  if (data=='a') GPIO_SetBits(GPIOA, GPIO_Pin_0);
  if (data=='b') GPIO_ResetBits(GPIOA, GPIO_Pin_0);
}
}
```

First of all we enable clocks for port A and USART1, without clock they won't work. Then we configure two pins we are going to use. Pin 0 is used to drive LED so it is configured as push-pull. On the other hand pin 10 is serial input so it is configured as floating input. Speed of both pins is set to 2 MHz because we don't need more. As we reuse the same structure we don't have to set this value each time. The next thing is configuration of USART1. We set the speed to 9600 bps, data length to 8 and width of stop bit to 1. There is no flow control or parity check. The USART's receiver is enabled. When these parameter as set we enable interrupt on receive. The next thing is to enable USART1. As you see this is not enough. We have configured USART1 to generate interrupts but we have to enable it with NVIC (Nested Vectored Interrupt Controller) too. STM32 offers advanced system of interrupt priorities. We have priority groups and subpriorities. The priority group is selected with *NVIC_PriorityGroupConfig()* function. Here we select group 0 which means highest priority. Then we enable interrupt for USART1 with subpriority 0. Now configuration is finished and main program can enter an infinite loop. When a byte is received then an interrupt is generated and is handled by *USART1_IRQHandler()*. Received byte is fetched using *USART_ReceiveData()* and then compared against 'a' and 'b' characters. If correct byte is received then pin 0 of port A is set to zero or one which causes LED to be turned off or on.

6.1.4. PC code

Writing code for RS-232 is already described in chapters devoted to serial port programming, therefore the sample code should be very easy to understand.

```
using System;
using System.Windows.Forms;
using System.IO.Ports;

namespace SerialSample1
{
    public partial class Form1 : Form
    {
        SerialPort ComPort = new SerialPort("COM1");
        public Form1()
        {
            InitializeComponent();
            if (ComPort.IsOpen) ComPort.Close();
            ComPort.BaudRate = 9600;
            ComPort.Parity = Parity.None;
            ComPort.StopBits = StopBits.One;
            ComPort.DataBits = 8;
            ComPort.Handshake = Handshake.None;
            ComPort.Open();
        }

        private void btnOn_Click(object sender, EventArgs e)
        {
            ComPort.Write("a");
        }

        private void btnOff_Click(object sender, EventArgs e)
        {
            ComPort.Write("b");
        }
    }
}
```

This is the source code of a standard Windows Forms window having two buttons: *btnOn* and *btnOff*. When the program starts, window object is instantiated from standard *main()* function, generated by Visual Studio. When the window is created, the *ComPort* object is initialized for first serial port in the system (COM1). If you want you can use another of available ports. Later the constructor of form object is called. It sets parameters of transmission and opens the serial port. The rest of the code is handlers for clicking on the two buttons. One of them is responsible for sending the 'a' character and the other one for sending the 'b' character. This way you can set state of the LED by clicking appropriate button.

As it was said in the chapter about AVR code, you can come up with a different protocol, e.g. state of a microcontroller's pin can be toggled every time a byte is received. But this may be realized also on the PC side as well, without modifying original microcontroller code. You just need a variable which holds last sent byte and when the button is clicked you need to send the other value: send 'b' if the last sent byte was 'a' and vice versa.

6.2. Receiving data with polling

In previous example the program on the microcontroller was using interrupts. That means that normal program execution was terminated to process incoming data and then was resumed. This is quite good approach because data can be analyzed as soon as they have been sent by the other device and the program can perform other tasks when waiting for data. There is however also a different approach, called active waiting. It is realized by constantly polling the state of USART.

6.2.1. Hardware configuration

The circuit remains exactly the same, for both AVR and STM32. The only change with respect to previous example is in software.

6.2.2. AVR code

Here is a code of the same functionality as in previous example but implemented using USART polling.

```
#include <avr/io.h>

void init_UART (void)
{
    UBRRL = 103; //9600 bps @ 16 MHz crystal
    UCSRB = _BV(RXEN); //EX enable
    UCSRC = _BV(URSEL)|_BV(UCSZ1)|_BV(UCSZ0); //n,8,1
}

int main() {
    init_UART();
    DDRA = _BV(DDA0);
    for(;;) {
        while(!(UCSRA & _BV(RXC)));
        uint8_t data = UDR;
        if (data=='a') {
            PORTA |= _BV(PINA0);
        }
        if (data=='b') {
            PORTA &= ~_BV(PINA0);
```

```
                }
        }
}
```

The code is similar but differs in lack of interrupt handling. We don't set the *RXCIE* bit in the *UCSRB* register so we don't enable interrupts for USART and we don't call the *sei()* function so interrupts are not enabled globally. Instead a new code is inserted into the infinite loop created using for loop. The active waiting is implemented by checking state of the *RXC* bit in the *UCSRA* register. The while loop iterates constantly until this bit is set. Hence the program waits until some data are received and *RXC* is set to 1. When it happens, data are read from the *UDR* register as in the first example. Following part of code is also the same, state of pin 0 of port A is set according to received value.

Although the program is very simple and doesn't execute other tasks than waiting for data and changing LED state, it's clearly visible that the execution is halted until data come.

6.2.3. STM32 code

In case of ARM the code is also similar to previous example.

```
#include "stm32f10x.h"

int main(void)
{

  RCC_APB2PeriphClockCmd(RCC_APB2Periph_GPIOA |
RCC_APB2Periph_USART1, ENABLE);

  GPIO_InitTypeDef GPIO_InitStructure;
  GPIO_InitStructure.GPIO_Pin = GPIO_Pin_10;
  GPIO_InitStructure.GPIO_Mode = GPIO_Mode_IN_FLOATING;
  GPIO_InitStructure.GPIO_Speed = GPIO_Speed_2MHz;
  GPIO_Init(GPIOA, &GPIO_InitStructure);

  GPIO_InitStructure.GPIO_Pin = GPIO_Pin_0;
  GPIO_InitStructure.GPIO_Mode = GPIO_Mode_Out_PP;
```

```
GPIO_Init(GPIOA, &GPIO_InitStructure);

USART_InitTypeDef USART_InitStructure;
USART_InitStructure.USART_BaudRate = 9600;
USART_InitStructure.USART_WordLength = USART_WordLength_8b;
USART_InitStructure.USART_StopBits = USART_StopBits_1;
USART_InitStructure.USART_Parity = USART_Parity_No;
USART_InitStructure.USART_HardwareFlowControl =
USART_HardwareFlowControl_None;
USART_InitStructure.USART_Mode = USART_Mode_Rx;
USART_Init(USART1, &USART_InitStructure);
USART_Cmd(USART1, ENABLE);

while (1) {
  while(USART_GetFlagStatus(USART1, USART_FLAG_RXNE) ==
RESET);
  char data = USART_ReceiveData(USART1);
  if (data=='a') GPIO_SetBits(GPIOA, GPIO_Pin_0);
  if (data=='b') GPIO_ResetBits(GPIOA, GPIO_Pin_0);
  }

}
```

The clock for port A and USART1 is configured as previously. Both peripherals are also configured the same way. The difference is that we don't enable interrupt for USART1 and we don't configure NVIC as well. There is no interrupt handler but in the infinite loop we wait for *RXNE* flag to be set. When it is set we get received byte and set state of pin 0 according value of this byte.

6.3.   *Reading button state*

This time we are considering opposite example: sending data from a microcontroller to a PC. Various kinds of data can be sent, coming from internal calculations or from external devices. Let's consider reading state of a button and sending it to a PC.

6.3.1. Hardware configuration

In this exercise we are going to replace the LED by a button which is going to be our simple input device. Again there are two styles of connection possible, the button can be connected to port pin from one side and to either VCC or GND on the other side. When connected to VCC the button will be generating logical 1 when pressed, logical 0 if connected to GND. Generally it's up to you, just adapt program's logic accordingly. What needs more attentions is a case when the button is not pressed. What logical level do we get then? Well, it depends. When the pin is truly unconnected, its state is undefined and is practically random. It can change randomly according to electrical field and electromagnetic noise. Thus input pins should not be left unconnected. To overcome the problem pull up resistors are used. Such a resistor is connected between a pin and VCC or GND, depending what logic level we would like to get. This way the input is immune to noise and its logic level is set by the resistor. When the button is pressed it enforces opposite level. It happens because the resistance of the pressed button is much lower than of the resistor.

We are using the general term of pull up resistor but it can be actually a pull up or pull down resistor depending whether it's connected to VCC or GND respectively. Pull up resistors are much more common however thus we can call both kinds as pull up resistors.

Discussing the problem of undefined state we were referring to truly unconnected pin. This term was used because many integrated circuits have in fact a built in pull up resistor. Such a resistor does its job and removes need for external one. Thus we can connect just a button alone. The ATmega32 microcontroller belongs to these chips but you must be aware of one thing: the pull up resistor is configurable, it can be internally connected or not. If it is not connected you have to provide an external pull up resistor.

The configuration for ATmega32 is made by write to port's output register. Normally we write to this register to set particular logic states on port's pins. However now our pin is configured as input so we are not changing its voltage. Instead we can turn on the internal pull up resistor. So if you programmatically enable the internal pull up resistor you don't need to attach the external one. The pull up resistors enabled for particular pins can be globally disabled by setting the *PUD* bit in the *SFIOR* register to one. Details about ATmega32 I/O ports can be found in the datasheet of this microcontroller.

In case of STM32 the pin we use for button is also configured by appropriate register. However we use the STM32 Standard Peripheral Library so it is not directly visible. There is just *GPIO_InitStructure.GPIO_Mode* field which is set to *GPIO_Mode_IPU*.

The ZL15AVR development board has four microswitches. They are connected from the GND side and have pull up resistors on the VCC side. Thus you don't have to care about pull up resistors. Just connect any one of them to a free pin of one of microcontroller's ports. In our example we are using pin 1 of port A. ZL30ARM has the same set microswitches and in example code we use pin 1 of port A as well.

Serial port and microcontrollers

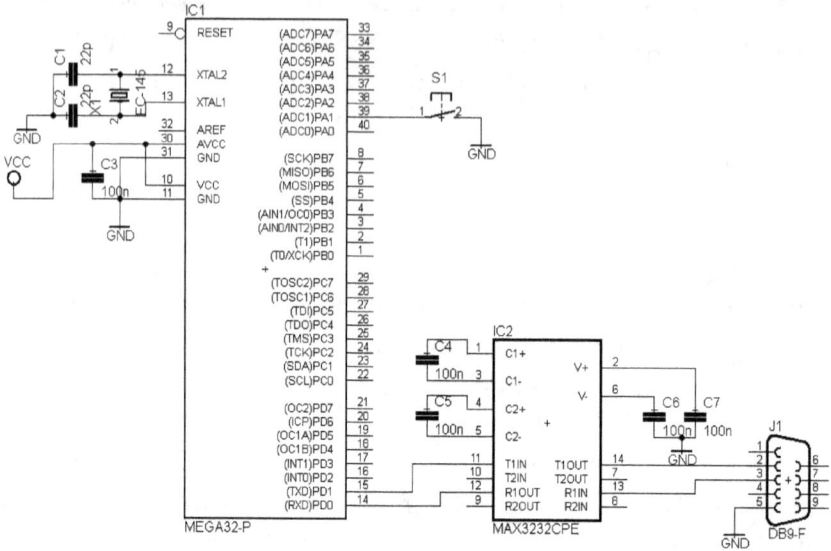

Fig. 6.3.1.1. Circuit with ATmega32.

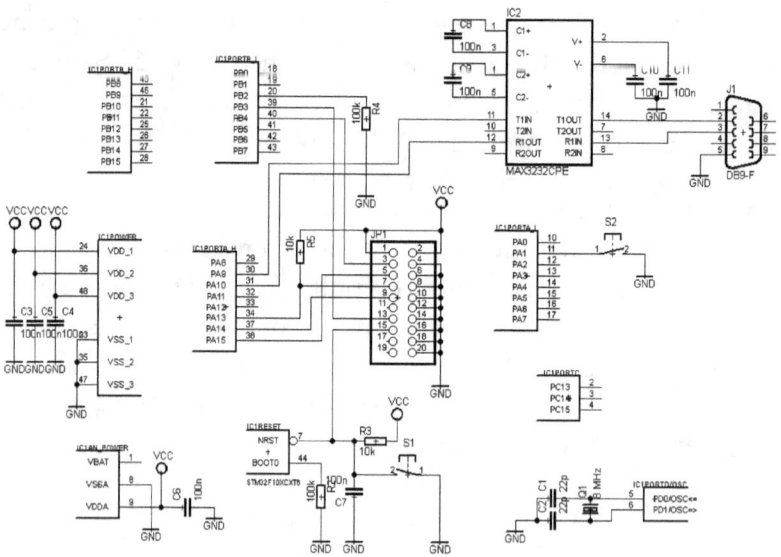

Fig. 6.3.1.2. Circuit with STM32F103CBT6.

6.3.2. AVR code

Here is a sample source code for our exercise:

```c
#include <avr/io.h>
#define F_CPU 16000000U
#include <util/delay.h>

void init_UART (void)
{
        UBRRL = 103; //9600 bps @ 16 MHz crystal
        UCSRB = _BV(TXEN); //TX enable
        UCSRC = _BV(URSEL)|_BV(UCSZ1)|_BV(UCSZ0); //n,8,1
}

int main() {
        uint8_t state;
        uint8_t stateOld = 3;
        PORTA |= _BV(PINA1);
        init_UART();
        for(;;) {
                state = PINA & _BV(PINA1);
                if (state!=stateOld) {
                        stateOld = state;
                        while(!(UCSRA & _BV(UDRE)));
                        if(state==0) UDR = 'P'; else UDR = 'N';
                        _delay_ms(50);
                }
        }
}
```

Looking back at previous example this one shouldn't be hard to understand. In USART initialization function we are enabling the transmitter part of USART instead of receiver. Of course you can have enabled both. Interrupts are not used. Pin 1 of port A works as input but with pull up resistor enabled. If you have an external pull up resistor connected you can omit this line.

In the main loop we are checking state of the pin and comparing with previous state. If the current state is different than the old one, we are sending information about this fact. To do this in the first order we wait for previous transmission to be finished by checking UDRE bit in *UCSRA* register. When *UDRE* (USART Data Register Empty) is set to 1 we can proceed. According to state of the pin the 'P' or 'N' character is sent. P stands for pressed, N for not pressed. This applies for the button connected at GND side, in this case the *state* variable is 0 when the button is pressed. When the button is not pressed the *state* variable has value 2 because we are using pin 1 ($2^1 = 2$). In the code we are just using the *else* clause, not checking particular value.

The data to be sent are written to the *UDR* register. Microcontroller takes this value and sends it through its TXD pin (pin 15). This is done by chip's hardware and we don't have to take care very much about it. We just have to check *UDRE* if we want to send another data.

You might be curious why the delay is introduced. Remove the call to *_delay_ms()* and look how the program works, use PuTTY or other terminal application. Have you noticed something strange? Sometimes when you press or release the button you can see that more than one character was transmitted, usually three. How does this happen? The problem is a result of the fact, that when you press or release a button the contacts doesn't touch cleanly but they bounce for a very short period of time. When this time is longer than the time needed to transmit one byte over RS-232 link then another transmission occurs because the contacts are still bouncing. There are several methods to solve this problem. One of them is a short delay, here it's 50 milliseconds. We are just waiting for bounces to stop and then proceed with checking state of the pin. The other solution is to add a capacitor in parallel with a button. The value of such a capacitor depends on pull up resistor and durations of bounces. The duration depends on mechanical construction of a switch and can vary from less than single microseconds to more than 100 milliseconds. Usually we deal with single milliseconds, thus 50 milliseconds used in the example should be enough. For a 10 kOhm pull up resistor a few nanofarads capacitor should be enough. In case of our example, where we are just sending information about state of a button the problem of bouncing may be not big but there are situation where this can be a major issue, e.g. when button causes an interrupt. Therefore take care about mechanical switches connected to your microcontroller.

The last thing to explain is initialization of the *stateOld* variable. Why is it set to 3? Well, the intention was to send the state of the button just after program starts so the other device doesn't have to wait for the button to be pressed or released to know what the state is. It gets the information as soon as the microcontroller is turned on. To do this we are simulating a button press/release by setting *state* and *stateOld* to different values. In this situation the current state is read and appropriate information is sent. Where the value 3 comes from? This value must be different than possible value read from pin. As we use pin 1, the possible values are 0 (button pressed) or 2 (button released). Generally it can be 0 or power of 2 depending on pin number (1 for pin 0, 2 for pin 1, 4 for pin 2 and so on). 3 is a safe number because it's not power of two so the *state* variable will never have such value, no matter which pin we use. It could be 3 only in case we are reading two pins (0 and 1) at the same time and both have value 1 ($1*2^1+1*2^0$).

6.3.3. STM32 code

Here is the code for STM32:

```
#include "stm32f10x.h"

void Delay(unsigned int delay)
{
  for(;delay>0;delay--);
}

int main(void)
{

    RCC_APB2PeriphClockCmd(RCC_APB2Periph_GPIOA |
RCC_APB2Periph_USART1, ENABLE);
```

```
    GPIO_InitTypeDef GPIO_InitStructure;
    GPIO_InitStructure.GPIO_Pin = GPIO_Pin_9;
    GPIO_InitStructure.GPIO_Speed = GPIO_Speed_2MHz;
    GPIO_InitStructure.GPIO_Mode = GPIO_Mode_AF_PP;
    GPIO_Init(GPIOA, &GPIO_InitStructure);

    GPIO_InitStructure.GPIO_Pin = GPIO_Pin_1;
    GPIO_InitStructure.GPIO_Mode = GPIO_Mode_IPU;
    GPIO_Init(GPIOA, &GPIO_InitStructure);

    USART_InitTypeDef USART_InitStructure;
    USART_InitStructure.USART_BaudRate = 9600;
    USART_InitStructure.USART_WordLength =
USART_WordLength_8b;
    USART_InitStructure.USART_StopBits = USART_StopBits_1;
    USART_InitStructure.USART_Parity = USART_Parity_No;
    USART_InitStructure.USART_HardwareFlowControl =
USART_HardwareFlowControl_None;
    USART_InitStructure.USART_Mode = USART_Mode_Tx;
    USART_Init(USART1, &USART_InitStructure);
    USART_Cmd(USART1, ENABLE);

    uint8_t stateOld = 3;

    while (1) {
            uint8_t state = GPIO_ReadInputDataBit(GPIOA,
GPIO_Pin_1);
            if (state != stateOld) {
                    stateOld = state;
                    while(USART_GetFlagStatus(USART1,
USART_FLAG_TXE) == RESET);
                    if (state==0) USART_SendData(USART1, 'P');
else USART_SendData(USART1, 'N');
                    Delay(100000);
            }
    }

}
```

This time we configure USART1 as transmitter, thus *USART_Mode* field is set to *USART_Mode_Tx*. Pin 9 of port A, which is USART1's

TX pin, is configured as push-pull output. AF states for alternate function and means that this pin doesn't work as general purpose I/O but works with peripheral circuit. State of the button is read using pin 1 so it is configured as input with pull-up resistor. State of the pin is read using *GPIO_ReadInputDataBit()* function. If it is different then we send a character according to current state: 'P' for zero and 'N' for '. To remove effect of bouncing we introduce 50 ms delay. It is realized using previously described function *Delay()*. The code checks state of USART1's TXE flag. It is actually not necessary because of the delay but it serves as a reminder that you shouldn't try to send a byte until transmission the previous one is finished.

6.3.4. PC code

In the example code we are displaying button's state. For this purpose simple Label control is used which is placed on a form. Let's look at form's code:

```
using System;
using System.IO.Ports;
using System.Windows.Forms;

namespace SerialSample3
{
    public partial class Form1 : Form
    {
        SerialPort ComPort = new SerialPort("COM1");
        String state = "";
        public Form1()
        {
            InitializeComponent();
            if (ComPort.IsOpen) ComPort.Close();
            ComPort.BaudRate = 9600;
            ComPort.Parity = Parity.None;
            ComPort.StopBits = StopBits.One;
            ComPort.DataBits = 8;
            ComPort.Handshake = Handshake.None;
            ComPort.DataReceived += OnSerialDataReceived;
            ComPort.Open();
        }
```

```
private void Form1_Load(object sender, EventArgs e)
{

}

private void OnSerialDataReceived(object sender,
SerialDataReceivedEventArgs args)
{
    int val = ComPort.ReadByte();
    if (val == 'P') state = "pressed"; else state =
"released";
    lblState.Invoke(new
EventHandler(lblState_DataReceived));
}

private void lblState_DataReceived(object sender,
EventArgs e)
{
    lblState.Text = "Button is " + state + ".";
}

    }
}
```

Initialization of serial port is done in the same manner as previously. The only difference is definition of handler routine for received data. This function, called *OnSerialDataReceived()*, reads received byte and sets the *state* variable accordingly. Then it calls label's *Invoke* method to execute its event handler. This handler sets text for the label. We are using the handler because the user interface runs in different thread than the code responsible for receiving data from serial port.

6.4.   *Two-way communication*

Until now we were focusing on communication in single direction, we were only receiving on only transmitting data. It's time to combine both functions. As an example we can consider tasks from previous examples: changing LED state and reading button state. It won't be hard because the two tasks are independent and whole thing is quite trivial. To make it more interesting we can add a simple improvement: checking button state on request. As you probably noticed, the problem with previous example was that when the program on a PC started, it didn't have the information from the microcontroller and the label displayed default text. It was updated when the state of the button was changed or when the microcontroller was turned on or reset. To fix this problem we are going to implement a way for a PC application to ask the microcontroller for current button state.

6.4.1. Hardware configuration

We are combining elements from previous example: a button and a LED. Here they are going to be connected like before. LED goes to pin 0 of port A and button to pin 1 of port A.

Serial port and microcontrollers

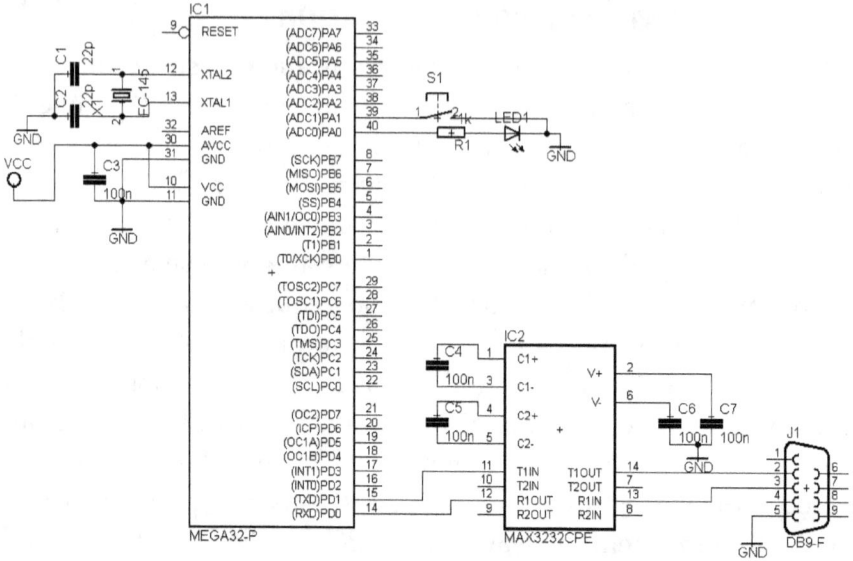

Fig. 6.4.1.1. Circuit with ATmega32.

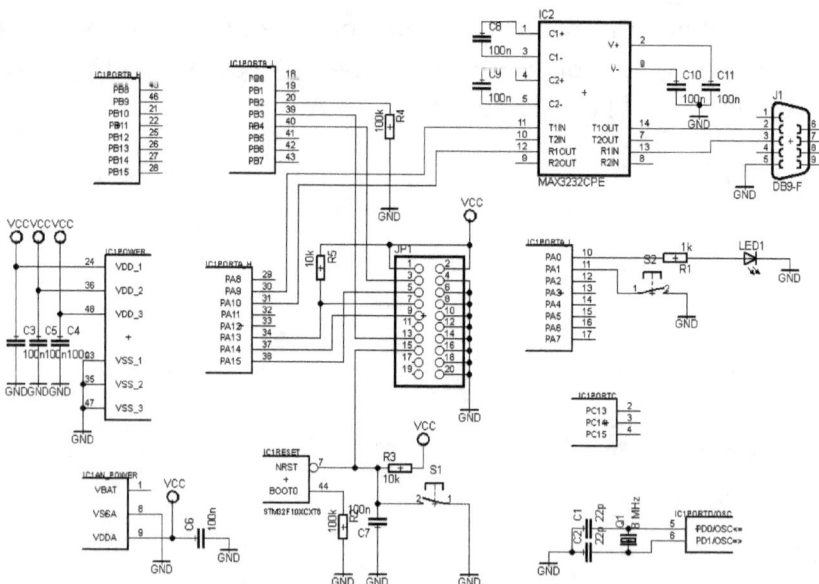

Fig. 6.4.1.2. Circuit with STM32F103CBT6.

6.4.2. AVR code

In the following code we are combining routines from previous exercises and adding code for handling new request from PC.

```c
#include <avr/io.h>
#include <avr/interrupt.h>
#define F_CPU 16000000U
#include <util/delay.h>

volatile uint8_t state;

void init_UART (void)
{
    UBRRL = 103; //9600 bps @ 16 MHz crystal
    UCSRB = _BV(RXEN)|_BV(TXEN)|_BV(RXCIE); //EX enable, RX
int enable
    UCSRC = _BV(URSEL)|_BV(UCSZ1)|_BV(UCSZ0); //n,8,1
}

ISR(USART_RXC_vect) {
    uint8_t data = UDR;
    if (data=='a') {
        PORTA |= _BV(PINA0);
    }
    if (data=='b') {
        PORTA &= ~_BV(PINA0);
    }
    if (data=='r') {
        while(!(UCSRA & _BV(UDRE)));
        if(state==0) UDR = 'P'; else UDR = 'N';
    }
}

int main() {
    uint8_t stateOld = 3;
    init_UART();
    sei();
    DDRA |= _BV(DDA0);
    PORTA |= _BV(PINA1);
    for(;;) {
        state = PINA & _BV(PINA1);
        if (state!=stateOld) {
```

```
                    stateOld = state;
                    while(!(UCSRA & _BV(UDRE)));
                    if(state==0) UDR = 'P'; else UDR = 'N';
                    _delay_ms(50);
            }
        }
}
```

In the USART initialization function both USART receiver and transmitter are enabled and interrupt for incoming data as well. In the USART interrupt handler a new branch was added. When the character 'r' is received the program sends state of the button. The *main()* function combines code from previous examples. It initializes USART, sets pin 0 of port A as output and enabled pull up resistor for pin 1 of port A. Then it sends button state if it changes.

An important change is new declaration of the *state* variable. It was moved outside of the *main()* function because it is used also in the USART interrupt handler. It is pretty obvious however and we should focus on another difference: addition of the *volatile* keyword. It tells the compiler not to optimize the code which accesses this variable. During optimization compiler modifies the source code to make it faster and smaller. It generally works really well but sometimes compiler can make a mistake. One of the most frequent reasons is a case when the compiler assumes that throughout some part of code a variable won't be changed. It can then for example remove parts of code executed for different values to make the code smaller. Or it can modify a code to use a register instead of piece of RAM to make access faster. If the variable is changed unexpectedly the new value will be ignored. A typical situation is when we use interrupts and it happens in our example. In the USART interrupt handle we are reading the *state* variable. Or, to be more precise, we want to read. When optimization is disabled, everything works as expected. But if you enable optimization you will notice that the read value is always zero, no matter whether the button is pressed or not.

Somehow during optimization compiler loses information that the variable is used in two places: interrupt handler and *main()* function. For the handler it assumes that the variable wasn't ever modified and has always the initial value. And if you don't specify the initial value in declaration compiler sets the value to zero. This is why our program would always send 'P' as a response to 'r', even if the button is not pressed. If we add the *volatile* keyword we tell the compiler that the variable can be unexpectedly changed and it should disable optimization for it. Then everything should work correctly.

6.4.3. STM32 code

Presented code is a combination of previous examples and actually doesn't contain anything new.

```
#include "stm32f10x.h"

void Delay(volatile unsigned int delay)
{
  for(;delay>0;delay--);
}

int main(void)
{

        RCC_APB2PeriphClockCmd(RCC_APB2Periph_GPIOA |
RCC_APB2Periph_USART1, ENABLE);

        GPIO_InitTypeDef GPIO_InitStructure;
        GPIO_InitStructure.GPIO_Pin = GPIO_Pin_9;
        GPIO_InitStructure.GPIO_Speed = GPIO_Speed_2MHz;
        GPIO_InitStructure.GPIO_Mode = GPIO_Mode_AF_PP;
        GPIO_Init(GPIOA, &GPIO_InitStructure);

        GPIO_InitStructure.GPIO_Pin = GPIO_Pin_10;
        GPIO_InitStructure.GPIO_Mode = GPIO_Mode_IN_FLOATING;
        GPIO_Init(GPIOA, &GPIO_InitStructure);

        GPIO_InitStructure.GPIO_Pin = GPIO_Pin_0;
        GPIO_InitStructure.GPIO_Mode = GPIO_Mode_Out_PP;
        GPIO_Init(GPIOA, &GPIO_InitStructure);
```

```
    GPIO_InitStructure.GPIO_Pin = GPIO_Pin_1;
    GPIO_InitStructure.GPIO_Mode = GPIO_Mode_IPU;
    GPIO_Init(GPIOA, &GPIO_InitStructure);

    USART_InitTypeDef USART_InitStructure;
    USART_InitStructure.USART_BaudRate = 9600;
    USART_InitStructure.USART_WordLength =
USART_WordLength_8b;
    USART_InitStructure.USART_StopBits = USART_StopBits_1;
    USART_InitStructure.USART_Parity = USART_Parity_No;
    USART_InitStructure.USART_HardwareFlowControl =
USART_HardwareFlowControl_None;
    USART_InitStructure.USART_Mode = USART_Mode_Tx |
USART_Mode_Rx;
    USART_Init(USART1, &USART_InitStructure);
    USART_ITConfig(USART1, USART_IT_RXNE, ENABLE);
    USART_Cmd(USART1, ENABLE);

    NVIC_PriorityGroupConfig(NVIC_PriorityGroup_0);
    NVIC_InitTypeDef NVIC_InitStructure;
    NVIC_InitStructure.NVIC_IRQChannel = USART1_IRQn;
    NVIC_InitStructure.NVIC_IRQChannelSubPriority = 0;
    NVIC_InitStructure.NVIC_IRQChannelCmd = ENABLE;
    NVIC_Init(&NVIC_InitStructure);

    uint8_t stateOld = 3;

    while (1) {
        uint8_t state = GPIO_ReadInputDataBit(GPIOA,
GPIO_Pin_1);
        if (state != stateOld) {
            stateOld = state;
            while(USART_GetFlagStatus(USART1,
USART_FLAG_TXE) == RESET);
            if (state==0) USART_SendData(USART1, 'P');
else USART_SendData(USART1, 'N');
            Delay(100000);
        }
    }

}
```

```
void USART1_IRQHandler(void)
{
    if (USART_GetITStatus(USART1, USART_IT_RXNE) != RESET)
    {
        char data = USART_ReceiveData(USART1);
        if (data=='a') GPIO_SetBits(GPIOA, GPIO_Pin_0);
        if (data=='b') GPIO_ResetBits(GPIOA,
GPIO_Pin_0);
    }
}
```

First of all we configure clocks and I/O pins. USART1 pins are configured in a standard way. Pin 0 is connected to switch so it is configured as input with pull-up. Pin 1 is connected to LED so it works as an output. USART1 is configured to work as both receiver and transmitter. Interrupt for incoming data is enabled and configured with NVIC. Parts of code responsible for sending and receiving data are as in previous examples. In the *Delay()* function we use the value 600 000 as the microcontroller works at 72 MHz and we want to have 50 ms delay.

6.4.4. PC code

The code of the PC application combines parts of code from previous example and is very easy to understand. Here is how it looks like:

```
using System;
using System.Windows.Forms;
using System.IO.Ports;

namespace SerialSample4
{
    public partial class Form1 : Form
    {
        SerialPort ComPort = new SerialPort("COM1");
        String state = "";
        public Form1()
        {
            InitializeComponent();
            if (ComPort.IsOpen) ComPort.Close();
            ComPort.BaudRate = 9600;
```

```
            ComPort.Parity = Parity.None;
            ComPort.StopBits = StopBits.One;
            ComPort.DataBits = 8;
            ComPort.Handshake = Handshake.None;
            ComPort.DataReceived += OnSerialDataReceived;
            ComPort.Open();
        }

        private void OnSerialDataReceived(object sender,
SerialDataReceivedEventArgs args)
        {
            int val = ComPort.ReadByte();
            if (val == 'P') state = "pressed"; else state =
"released";
            lblState.Invoke(new
EventHandler(lblState_DataReceived));
        }

        private void lblState_DataReceived(object sender,
EventArgs e)
        {
            lblState.Text = "Button is " + state + ".";
        }

        private void btnOn_Click(object sender, EventArgs e)
        {
            ComPort.Write("a");
        }

        private void btnOff_Click(object sender, EventArgs e)
        {
            ComPort.Write("b");
        }

        private void Form1_Load(object sender, EventArgs e)
        {
            ComPort.Write("r");
        }

    }
}
```

There is not much we can say about above code, almost all parts were discussed earlier. The only modification is addition of sending the 'r' character when the program's window shows up. The connected microcontroller responds with state of the button and this way when we start the application we always see current state of the button.

6.5. Reading multibyte data – a voltmeter

Till now we were transmitting single bytes, now let's consider data having more than one byte. As an example let's consider reading a value from analog-to-digital converter (ADC) which both ATMega32 and STM32F103CBT6 have built in. We can use it as a simple digital voltmeter. In this chapter we are considering a microcontroller as a transmitter of multibyte data. Receiving such data by a microcontroller is talked over in one of following chapters.

6.5.1. Hardware configuration

In this example we are going to measure a voltage which means that we have to connect some voltage source to our microcontroller. You can use various voltage sources but of course in the range safe for your chip. It's a good idea to take a potentiometer and plug it between GND and VCC, then the slider terminal connect to a microcontroller. This way you can set arbitrary voltage in the range from 0 volts to VCC. ATmega32 has 8 ADC channels (pins 33 – 40). In our example we use ADC1 (pin 39). STM32F103CBT6 has two ADC two ADCs, each one has 16 channels. In our STM32 example we use channel 2 of ADC1 (pin PA2).

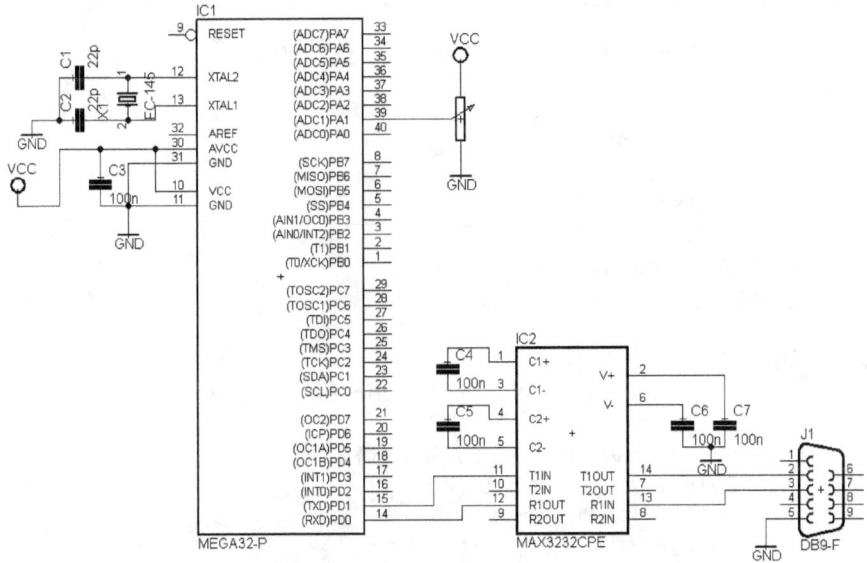

Fig. 6.5.1.1. Circuit with ATmega32.

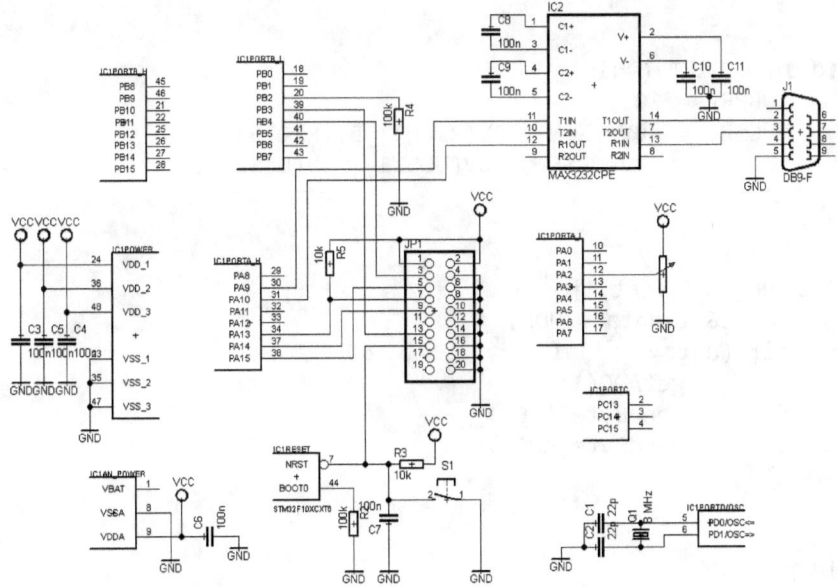

Fig. 6.5.1.2. Circuit with STM32F103CBT6.

6.5.2. AVR code

Here is sample source code for our microcontroller:

```c
#include <avr/interrupt.h>
#include <avr/io.h>

void sendByte(char c);
void init_UART(void);
void init_ADC(void);
void getADC();

void init_ADC(void) {
    // Internal 2.56V voltage source as reference voltage
for ADC
    ADMUX |= _BV(REFS0)|_BV(REFS1);
    // Select ADC channel (ADC1)
    ADMUX |= _BV(MUX0);
    // Enable ADC
    ADCSRA |= _BV(ADEN);
    // Set ADC clock prescaler
    ADCSRA |= _BV(ADPS0)|_BV(ADPS1)|_BV(ADPS2);
}

void init_UART (void) {
    UBRRL = 103;
    UCSRB = _BV(TXEN)|_BV(RXEN)|_BV(RXCIE);
    UCSRC = _BV(URSEL)|_BV(UCSZ1)|_BV(UCSZ0);
    sei();
}

ISR(USART_RXC_vect) {
    uint8_t data = UDR;
    if (data=='x') {
        getADC();
        sendByte(ADCL);
        sendByte(ADCH);
    }
}

void getADC() {
    ADCSRA |= _BV(ADSC);
    while(ADCSRA & _BV(ADSC)) {};
```

```
}

int main(void)
{
        init_UART();
        init_ADC();
        while(1);
}

void sendByte(char c) {
        while(!(UCSRA & _BV(UDRE)));
        UDR = c;
}
```

The source code is divided into several functions which are declared on the beginning. *init_ADC()* performs analog-to-digital converter initialization. First of all we select reference voltage source. You can choose from three of them: VCC, AREF and internal 2.56 V reference source. The ADC in ATmega32 has 10-bit resolution so it can give values in the range from 0 to 1023 (0000h to 03FFh). The reference voltage is the voltage at which the value returned by ADC is maximal. In our example we chose the 2.56 V internal source which gives as the resolution of 2.56 V / 1023 = 2.5 mV. So for example for 100 mV the output value would be 40 (100 / 2.5).

Be aware that internal 2.56V reference voltage source doesn't have good precision. If you read ATmega32 datasheet you won't find this information in the chapter devoted to ADC. You have to jump to *Electrical Characteristics* chapter and scroll to ADC. There you can read that internal reference voltage can vary from 2.3 to 2.7 volts. Therefore you should compare the value displayed by the PC application with the value you read on your voltmeter and modify the PC application if the difference is big enough that can't be ignored.

The next thing is to select ADC channel. In ATmega32 there are 8 ADC channels and you can pick anyone of them. This way you can measure 8 different voltages. There is only one ADC in the microcontroller however, which means that you have to read a value from one channel, switch to the second one and then make the second measurement. You can select the channel by setting appropriate bits in *ADMUX* register. Bits *MUX4* and *MUX3* must be set to zero, bits *MUX2..0* select ADC channel. The *MUX4* and *MUX3* enable differential measurements, where the voltage difference between two channels is measured instead of absolute values. Additionally it can be done with different gains (1, 10, 200). You are very much encouraged to experiment with differential mode but here we are sticking with absolute measurements.

By setting the *ADEN* byte in the *ADCSRA* register we turn on the ADC. By setting the *ADPS0..2* bits we set the prescaler for ADC clock. It determines how many times the ADC clock will be slower than the system clock. The datasheet states that default, the successive approximation circuitry requires an input clock frequency between 50 kHz and 200 kHz to get maximum resolution. If we have 16 MHz crystal and prescaler set to 128 it gives us the frequency of 125 kHz which is located inside mentioned interval. Thus all *ADPS2..0* bits in *ADCSRA* are set to 1.

Reading data from ADC is done in the *getADC()* function. The process of conversion is started by setting the *ADSC* bit in the *ADCSRA* register. Then the program waits for the conversation to be completed. It is done by observing the bit, when the conversation is finished it is changed back to 0. The result for analog-to-digital conversion is stored in two registers: *ADCL* and *ADCH*. By default *ADCH* holds 2 more significant bits and ADCL 8 less significant bits. We say that the data are right-adjusted. If the *ADLAR* bit in the *ADMUX* register is set to 1 then data are left-adjusted: there are eight more significant bits in *ADCH* and two less significant in *ADCL*. In our example we use the default mode. It's convenient to have data left adjusted if you need 8-bit or lower resolution. In this case you can just take the value from *ADCH* and don't care about *ADCL*. If the data were right-adjusted you would have to take two bits from *ADCH* and some from *ADCL*.

The routine for USART initialization is a standard one, with both transmitter and receiver enabled and the interrupt for incoming data. Like previously the transmission parameters are: 9600 bps at 16 MHz crystal, 8 bits of data, one stop bit and no parity bits. The piece of code responsible for sending data over RS-232 has been moved to separate function *sendByte()*.

In the USART interrupt routine we are waiting for the 'x' character. When it is received we read a value from ADC. Then we send contents of *ADCL* and *ADCH*. This means that first we send less significant part of the value obtained from ADC and then the more significant. From the transmission point of view the order could be reversed, it doesn't matter which part is sent first and which second as long as the receiver knows which is which. But from the point of view of ADC it is not so simple. If you have some experience with microcontrollers you know that register often are not just simple pieces of memory and both reading and writing them can trigger some mechanisms. We were already discussing it when describing the UDR register of USART.

It happens that one piece of code reads *ADCH* and other one reads *ADCL*. This is how it is done in our example although they are next to each other. Generally it is possible to read *ADCH* and *ADCL* in different moments in time. It can happen that the end of analog-to-digital conversion happens between those two reads. In effect one read would belong to different conversion than the other. In our example there is no such problem because we trigger the conversion and wait for it to finish. But in general the problem can occur. To prevent this, a simple logic was introduced by the microcontroller's designers: as long as *ADCH* is not read the ADC data register is blocked and not updated by further conversions. This has two consequences: 1) if the data are left-adjusted and you need only 8-bits or less you can read just *ADCH* and everything is OK; 2) if you would like to read *ADCH* and *ADCL* you have to read *ADCL* first and then *ADCH*. In our example we deal with the second case so we are sending *ADCL* first. If we wanted to send *ADCH* first we would need to copy *ADCL* to a variable, send *ADCH* and the saved value of *ADCL*.

6.5.3. STM32 code

```
#include "stm32f10x.h"

void init_USART();
void init_ADC();

int main(void)
{
        init_USART();
        init_ADC();

        while(1);

}

void USART1_IRQHandler(void)
{
        if (USART_GetITStatus(USART1, USART_IT_RXNE) != RESET)
        {
                char data = USART_ReceiveData(USART1);
                if (data=='x') {
                        uint16_t adcVal =
ADC_GetConversionValue(ADC1);
                        adcVal >>= 2;
                        while(USART_GetFlagStatus(USART1,
USART_FLAG_TXE) == RESET);
                        USART_SendData(USART1, adcVal & 0xff);
                        while(USART_GetFlagStatus(USART1,
USART_FLAG_TXE) == RESET);
                        USART_SendData(USART1, adcVal >> 8);
                }
        }
}
```

```
void init_ADC() {
        RCC_APB2PeriphClockCmd(RCC_APB2Periph_ADC1, ENABLE);

        GPIO_InitTypeDef GPIO_InitStructure;
        GPIO_InitStructure.GPIO_Pin = GPIO_Pin_2;
        GPIO_InitStructure.GPIO_Mode = GPIO_Mode_AIN;
        GPIO_InitStructure.GPIO_Speed = GPIO_Speed_2MHz;
        GPIO_Init(GPIOA, &GPIO_InitStructure);

        ADC_InitTypeDef ADC_InitStructure;
        ADC_InitStructure.ADC_Mode = ADC_Mode_Independent;
        ADC_InitStructure.ADC_ScanConvMode = DISABLE;
        ADC_InitStructure.ADC_ContinuousConvMode = ENABLE;
        ADC_InitStructure.ADC_ExternalTrigConv =
ADC_ExternalTrigConv_None;
        ADC_InitStructure.ADC_DataAlign = ADC_DataAlign_Right;
        ADC_InitStructure.ADC_NbrOfChannel = 1;
        ADC_Init(ADC1, &ADC_InitStructure);
        ADC_RegularChannelConfig(ADC1, ADC_Channel_2, 1,
ADC_SampleTime_71Cycles5);
        ADC_Cmd(ADC1, ENABLE);
        ADC_ResetCalibration(ADC1);
        while(ADC_GetResetCalibrationStatus(ADC1));
        ADC_StartCalibration(ADC1);
        while(ADC_GetCalibrationStatus(ADC1));
        ADC_SoftwareStartConvCmd(ADC1, ENABLE);
}

void init_USART(){

        RCC_APB2PeriphClockCmd(RCC_APB2Periph_GPIOA |
RCC_APB2Periph_USART1, ENABLE);

        GPIO_InitTypeDef GPIO_InitStructure;
        GPIO_InitStructure.GPIO_Pin = GPIO_Pin_9;
        GPIO_InitStructure.GPIO_Speed = GPIO_Speed_2MHz;
        GPIO_InitStructure.GPIO_Mode = GPIO_Mode_AF_PP;
        GPIO_Init(GPIOA, &GPIO_InitStructure);

        GPIO_InitStructure.GPIO_Pin = GPIO_Pin_10;
        GPIO_InitStructure.GPIO_Mode = GPIO_Mode_IN_FLOATING;
        GPIO_Init(GPIOA, &GPIO_InitStructure);
```

```
      USART_InitTypeDef USART_InitStructure;
      USART_InitStructure.USART_BaudRate = 9600;
      USART_InitStructure.USART_WordLength =
USART_WordLength_8b;
      USART_InitStructure.USART_StopBits = USART_StopBits_1;
      USART_InitStructure.USART_Parity = USART_Parity_No;
      USART_InitStructure.USART_HardwareFlowControl =
USART_HardwareFlowControl_None;
      USART_InitStructure.USART_Mode = USART_Mode_Tx |
USART_Mode_Rx;
      USART_Init(USART1, &USART_InitStructure);
      USART_ITConfig(USART1, USART_IT_RXNE, ENABLE);
      USART_Cmd(USART1, ENABLE);

      NVIC_PriorityGroupConfig(NVIC_PriorityGroup_0);
      NVIC_InitTypeDef NVIC_InitStructure;
      NVIC_InitStructure.NVIC_IRQChannel = USART1_IRQn;
      NVIC_InitStructure.NVIC_IRQChannelSubPriority = 0;
      NVIC_InitStructure.NVIC_IRQChannelCmd = ENABLE;
      NVIC_Init(&NVIC_InitStructure);
}
```

USART in this example is configured in standard way and works as a receiver and transmitter. Its configuration has been moved to separate function for better readability. Analog-to-digital converter needs more attention. Here we use ADC1. It has several channels, we use channel 2. The pin serving as an input for this channel is pin 2 of port A. It is going to be used as an analog input so we configure it with *GPIO_Mode_AIN*.

ADC1 works in the independent mode which means that it is not synchronized with other ADCs. With *ADC_ScanConvMode* set to DISABLE we are not using scan mode. In this mode ADC performs conversion on consecutive channels. We use only one channel so we don't need the scan mode. On the other hand we enable continuous conversion mode. With it we don't have to start conversion each time we want to read a value from ADC because the conversion is performed continuously. We don't use an external trigger for conversion. The next thing being configured is how data are aligned.

The ADC our chip is 12-bit which means there is some free space in 16-bit register and data can be aligned left or right. We use right align because it is more convenient. The field *NbrOfChannel* is set to 1 which means we configure one channel. The name of this field suggests that it is number of the channel being configured but in fact it determines how many channels are being configured.

With *ADC_RegularChannelConfig()* we set additional options for chosen channel. The first parameter is the ADC we configure and the second one is the channel. The third parameter is rank in the regular group sequencer. As we use only one channel this parameter is not important very much. The last parameter is sample time expressed in clock cycles. ADC performs a conversion in 12.5 cycles + programmed value. Thus if the last parameter is 1.5 then ADC has 14 cycles to perform conversion. If ADC works with 14 MHz clock then the speed of conversion is 1 MHz which means that it takes exactly 1 microsecond to obtain result of analog-to-digital conversion. That's maximal speed of ADC. Because the source of clock signal for ADC jest the APB2 bus you might sometimes have to introduce division of APB2 clock for ADC. This is done with *RCC_ADCCLKConfig()* function. For example if APB2 has speed of 56 MHz you need to divide it by four so you call *RCC_ADCCLKConfig()* with parameter *RCC_PCLK2_Div4*. In our case high speed of conversion is not necessary and we use *ADC_SampleTime_71Cycles5* value. This gives us 84 cycles. The main clock is 8 MHz because we didn't change default value. APB2 works with this frequency because no division for this bus was introduced as well. If we divide 8 MHz by 84 we get a little bit more than 95 kHz. That's how fast ADC is in our configuration.

When ADC1 is configured we enable it. What we have more to do is to perform calibration of the converter. We start with resetting calibration registers and then perform actual calibration. The last thing is to start conversion.

Serial port and microcontrollers

The code responsible for communication is simple. When a USART interrupt happens we check if there are new data available. If the character x was received we acquire result of A/D conversion and send its lower byte and higher byte. You can notice that we right-shift the read value. This is because the PC application expects 10-bit value and STM32 provides 12-bit ADC resolution. If you want to use 12-bit resolution remove the bit-shift and modify the PC application accordingly.

6.5.4. PC code

The C# code is short but interesting and shows one trick for Windows Vista and Windows 7 users.

```
using System;
using System.Windows.Forms;
using System.IO.Ports;

namespace SerialSample5
{
    public partial class Form1 : Form
    {
        SerialPort ComPort = new SerialPort("COM1");
        int val = 0;
        public Form1()
        {
            InitializeComponent();
            if (ComPort.IsOpen) ComPort.Close();
            ComPort.BaudRate = 9600;
            ComPort.Parity = Parity.None;
            ComPort.StopBits = StopBits.One;
            ComPort.DataBits = 8;
            ComPort.Handshake = Handshake.None;
            ComPort.DataReceived += OnSerialDataReceived;
            ComPort.Open();
        }

        private void OnSerialDataReceived(object sender,
SerialDataReceivedEventArgs args)
```

```
        {
            int lByte = ComPort.ReadByte();
            int hByte = ComPort.ReadByte();
            val = (hByte<<8) + lByte;
            lblValue.Invoke(new
EventHandler(lblValue_DataReceived));
        }

        private void lblValue_DataReceived(object sender,
EventArgs e)
        {
            lblValue.Text = val.ToString();
            progressBar1.Value = val;
            if (val > 0) progressBar1.Value = val - 1;
            progressBar1.Value = val;
            double valf = (double) val / 399.6;
lblVoltage.Text = valf.ToString("0.0000") + " V";
ComPort.Write("x");
        }

        private void Form1_Load(object sender, EventArgs e)
        {
            ComPort.Write("x");
        }
    }
}
```

The cod for serial port initialization is identical as previously. When the form loads we send the 'x' character to our microcontroller to get data from its ADC. That data is received in *OnSerialDataReceived()* function. First we read less significant byte, then the more significant. To combine them we are adding them after the bits in the more significant byte were shifted left by 8 positions. This makes these bits really more significant and places them in right position. In the end we are invoking event handler of one of the labels on the form, just like in previous examples.

In the event handler we set text of the *lblValue* label. We use *ToString()* to convert from *int* to *string*. Then we are updating the progress bar which serves here are nice data visualization object. As we turn the potentiometer we change how much the progress bar is filled. Setting of the progress bar value is done in a strange way: we are setting the value, then the value lower by one and again the right one. It is the trick mentioned on the beginning. It's a workaround for a "feature" Microsoft introduced in Windows Vista. When a value greater than previous one is set, the progress bar doesn't change its value immediately but progressively. Maybe it creates nice effect in setup wizards etc. but is annoying in or program which shows voltage which can change rapidly, eg. when you make fast potentiometer turn. The trick uses the fact that there is no delay when the value is changed to lower one. So the simple workaround is to set the right value, then lower by one and then again the right one. If you have Windows XP you can remove two lines which do the hack.

The next thing to do is to update the *lblVoltage* label which shows the voltage. To obtain the value expressed in volts we have to divide received value by 399.6. This comes from the fact that for 2.56 V we get the value of 1023 (03FFh). If different reference voltage was used then different value must be calculated. For example if the reference voltage is VCC = 5 V then the value would be 1023 / 5 = 204.6. The *val* variable is casted to *double* because without this we would get integer division so the *valf* variable could be only 0, 1 or 2 as the result of division. The resulting voltage is presented with 4 positions after decimal point. You can use more or less decimal points but must have in memory that ADC in ATmega32 is only 10-bit so it doesn't make sense to use many decimal positions.

Because we want to have current value the last thing the event handler does is to send the 'x' character so new data can be received. This way the PC application constantly asks the microcontroller for current value from ADC.

If the application is going to work with STM32, not AVR, remember that STM32 doesn't have internal 2.56 reference voltage and uses supply voltage as reference. Therefore you have to do some simple math and modify the application.

6.6. Reading continuous data stream

If you were thinking about our last example you could ask: why does the PC have to constantly ask for data? Couldn't the microcontroller send it on and on? The answer is yes, but some conditions must be met.

The most important condition is related to the RS-232 standard itself. The start of data frame is recognized by the start bit. This start bit must be preceded by a stop bit belonging to previous frame or by some period of inactivity on RS-232 line. In other case, when the receiver is turned on in the middle of transmitted frame, it can take one of transmitted bits as a start bit. As the result we would get garbage. Even when the true start bit comes it won't be recognized correctly. Thus there must be some time of silence on the wire. It can be realized in two similar ways and a third one, a little bit different:

- simply by adding a delay at least as long as transmission of one byte

- by transmitting a byte of value 255 (FFh) which consists solely of ones so doesn't introduce zero which could be incorrectly interpreted as start bits

- by setting stop bit length to 1.5; this however requires more advanced, intelligent USARTs so it's not a universal solution

This way we are left we the first two approaches. Now we have the problem of start bits almost solved. I say almost because the silence or FFh byte are synchronization points and until they occur the receiver gets garbage. This means that for a short time the receiver has to ignore incoming data. Thus the second approach is much better because we don't have to measure time of inactivity but simply wait for the FFh byte.

But again it's not the end of issues we can encounter. We have solved the problem on the low, RS-232 level but on the higher level situation can be still complicated. If we didn't decide to use delays but the FFh byte then a new problem emerges: we can no longer transmit arbitrary data because the value FFh is reserved. If you transmit some text data or other bytes lower than FFh then there is no problem. But in other case you need some method of data encapsulation so arbitrary data can by expressed as set of bytes from which none has the value FFh.

One of them, very simple by the way, is representing data as hexadecimal numbers. This way we are not transmitting bytes of any value but from the range 48 – 57 for characters '0' – '9' and 65 – 70 for characters 'A' – 'F'. For example to send a 16-bit number of value 45969d we have to send following bytes: FFh 42h 33h 39h 31h. The last four bytes are encoding characters: 'B', '3', '9' and '1'. This is because of the fact that 45969 in decimal notation is B391 in hexadecimal notation.

As you can see data encapsulation introduces some overhead. In our example we are sending five bytes to actually send only two bytes of data. There is one start byte and two bytes for each data byte. If you consider this a waste of bandwidth you can create better encoding scheme, utilizing smaller number of bytes. The algorithm presented above is just an example. Its advantage is that the data can be easily viewed with PuTTY or similar serial terminal application because of hexadecimal notation.

How can we send 16-bit value using less than 5 bytes? As you can guess there are plenty possibilities while 5 bytes is much more than 2 bytes. Let's consider 3-byte example. We would like to distinguish somehow between data bytes and the delimiter byte. For example we can assume that if the most significant bit is 0 then it's data byte and if it is 1 then we are dealing with the delimiter byte. As you can see we have only 14 bits left for data and we still have to transmit 2 remaining bits. Because we have 7 unused bits in the delimiter byte, there is no problem to use two of them for data. This way we are using three bytes for 16-bits of data (7+7+2). Thus from 24 bits we use 16 for data, 3 for byte type and 5 are left unused. As you can see we can make our encoding even more efficient, for example using 5 bytes for two 16-bit values. In this case 32-bits out of 40 are used for data, 5 for byte type and 3 are unused.

In case of ADC in ATmega32 we have to send in fact 10 bits, for example using 5 less significant bits in each of two bytes. Hence for example for value 678 (02A6h) we are going to send: FFh 15h 06h. Where these values come from? Let's express original value in binary. 02h is 00000010b and A6 is 10100110b. Now we create two groups of 5 bits taking 2 bits from the more significant byte and 3 bits from the less significant byte for the first group and remaining 5 bits for the second group: 10101 and 00110. To have bytes we need to add leading zeros: 00010101 and 00000110. This way we have bytes: 15h and 06h. OK, it's not complicated. But trying to take the exact same approach when writing code for a microcontroller can be tricky. The easier to implement algorithm will make use of bit shift instructions. Let's assume you have two bytes in the form of 16-bit variable. If you have two separate variables just copy the one which is going to be more significant and copy to 16-bit variable, perform left shift by 8 positions and add the second variable. Having 16-bit variable perform following steps:

- copy 16-bit variable to 8-bit variable, the one considered as less significant, the more significant 8-bits will be truncated so only less significant bits will be copied to the variable

- perform an AND bit operation on the 8-bit variable with the value 1Fh (00011111b) so you can get 5 bits

- right shift the 16-bit variable by 5 positions

- copy 16-bit variable to another 8-bit variable, the one considered more significant

Let's assume the 16-bit variable (uint16_t) is called *big* and the 8-bit variables (uint8_t) are called *less* and *more*. The sample source code can look as follows:

```
less = big;
less &= 0x1F;
big >>= 5;
more = big;
```

Above deliberation is intended to introduce some basic aspects of data encoding and present some problems which can be encountered. In practice we don't use RS-232 links to transmit big amounts of data with high speed so overhead of encapsulation is not big deal. The more important thing is reliability of used protocol. It should be also easy to analyze and implement.

As for the example for this chapter we are going to study the same problem: constantly reading and transmitting value from ADC. This time the PC application doesn't ask for data, the microcontroller sends data solely by itself. We'll see how to encode using hexadecimal numbers and decode them properly. We are also going to implement presented above encoding of 10-bit values.

6.6.1. Hardware configuration

In this example hardware remains the same as in previous example. If you wish you can add some modifications, e.g. LED turned on when value read from ADC exceeds some value. Of course you will need to add appropriate piece of code which should be easy for you at this moment.

6.6.2. AVR code

Let's look at sample code:

```c
#include <avr/io.h>
#include <stdio.h>

void sendByte(char c);
void init_UART(void);
void init_ADC(void);
uint16_t getADC();
void hexStr(char * buf, uint16_t val);

void init_ADC(void) {
        // Internal 2.56V voltage source as reference voltage
for ADC
        ADMUX |= _BV(REFS0)|_BV(REFS1);
        // Select ADC channel (ADC1)
        ADMUX |= _BV(MUX0);
        // Enable conversion
        ADCSRA |= _BV(ADEN);
        // Set ADC prescaler
        ADCSRA |= _BV(ADPS0)|_BV(ADPS1)|_BV(ADPS2);
}

void init_UART (void) {
        UBRRL = 103;
        UCSRB = _BV(TXEN);
        UCSRC = _BV(URSEL)|_BV(UCSZ1)|_BV(UCSZ0);
}

uint16_t getADC() {
        ADCSRA |= _BV(ADSC);
        while(ADCSRA & _BV(ADSC)) {};
        return ADCW;
}

int main(void)
{
        init_UART();
        init_ADC();
        char buf[5];
        while(1) {
```

```
                hexStr(buf, getADC());
                sendByte(0xff);
                for (int i=0; i<4; i++) sendByte(buf[i]);
        }
}

void sendByte(char c) {
        while(!(UCSRA & _BV(UDRE)));
        UDR = c;
}

void hexStr(char * buf, uint16_t val) {
        sprintf(buf, "%04X", val);
}
```

The function which configures ADC is exactly the same as in previous example. The procedure for reading ADC value has changed a little bit. Now it returns contents of ADCL and ADCH at once using ADCW register. In the USART configuration code the part for enabling receiver and interrupts has been removed as it is not necessary in this example. On the other hand there is no problem if the code remained the same as previously. Code for sending bytes remained unchanged.

The important addition is the *hexStr()* function which is responsible for converting 16-bit value into string of hexadecimal characters. It is done using popular C function called *sprintf()*. It takes a value and creates appropriately formatted string in provided buffer. In our example the formatting string is 04X which means that we want to have a hexadecimal number written using 4 characters, including leading zeros. It must be noted that *sprintf()* introduces huge overhead. With this function used whole program takes 1830 bytes while it took 290 bytes without it. Your results may vary depending on compiler version but the difference is huge anyway. In case of ATmega32 which has 32 kBytes of Flash memory it probably doesn't make a big deal, but if you are going to write code for some of the most simple, low cost chips, having 1 or 2 kBytes of Flash memory, it can be a big problem. In such a case you have to find other way for conversion to hex, looking at available libraries or writing own function.

As you can see, the *main()* function is pretty simple. In an infinite loop we are getting a value from ADC and converting the value to string. Then we send the FFh byte and the consecutive characters representing hexadecimal digits of measured value. The buffer is 5 bytes big but only 4 bytes are used. The fifth byte is allocated because *sprintf()* function places the null-character after the string. In other words it returns a null-terminated string. The null-characters is simply a byte of value 0 and marks end of string. In out example it doesn't matter because we are always interested in fixed number of characters and we don't need any ending marker. But still we need a place for the null-character. If the buffer was only 4 bytes large, the program would probably work correctly. There are no checks for out of bound array access. Simply a byte of memory located after the buffer would be overwritten. Is it dangerous? It depends whether there are some other variables and how a compiler places them in memory. It may happen that some variable is located just after the buffer and writing out of buffer bounds will cause this variable to be overwritten and this can lead to unexpected program behavior. Thus you have to pay attention at arrays sizes.

6.6.3. STM32 code

The modifications introduced in this example in relation to the previous one are analogous to the ones made in AVR code. We don't use interrupts so the code responsible for them has been removed. Instead we constantly send measured value. Like in the AVR code we use standard *sprint()* function to obtain string with hexadecimal value. The byte FFh is sent for synchronization purposes. The other change is that the two lines responsible for sending a byte have been moved to separate function which increases readability.

```c
#include "stm32f10x.h"
#include <stdio.h>

void init_USART();
void init_ADC();
void hexStr(char * buf, uint16_t val);
void sendByte(char c);

int main(void)
{
        init_USART();
        init_ADC();

        char buf[5];
        while(1) {
                uint16_t adcVal = ADC_GetConversionValue(ADC1);
                adcVal >>= 2;
                hexStr(buf, adcVal);
                sendByte(0xff);
                for (int i=0; i<4; i++) {
                        sendByte(buf[i]);
                }
        }

}

void hexStr(char * buf, uint16_t val) {
        sprintf(buf, "%04X", val);
}

void sendByte(char c) {
        while(USART_GetFlagStatus(USART1, USART_FLAG_TXE) ==
RESET);
        USART_SendData(USART1, c);

}

void init_ADC() {
        RCC_APB2PeriphClockCmd(RCC_APB2Periph_ADC1, ENABLE);

        GPIO_InitTypeDef GPIO_InitStructure;
        GPIO_InitStructure.GPIO_Pin = GPIO_Pin_2;
        GPIO_InitStructure.GPIO_Mode = GPIO_Mode_AIN;
```

```
      GPIO_InitStructure.GPIO_Speed = GPIO_Speed_2MHz;
      GPIO_Init(GPIOA, &GPIO_InitStructure);

      ADC_InitTypeDef ADC_InitStructure;
      ADC_InitStructure.ADC_Mode = ADC_Mode_Independent;
      ADC_InitStructure.ADC_ScanConvMode = DISABLE;
      ADC_InitStructure.ADC_ContinuousConvMode = ENABLE;
      ADC_InitStructure.ADC_ExternalTrigConv =
ADC_ExternalTrigConv_None;
      ADC_InitStructure.ADC_DataAlign = ADC_DataAlign_Right;
      ADC_InitStructure.ADC_NbrOfChannel = 1;
      ADC_Init(ADC1, &ADC_InitStructure);
      ADC_RegularChannelConfig(ADC1, ADC_Channel_2, 1,
ADC_SampleTime_71Cycles5);
      ADC_Cmd(ADC1, ENABLE);
      ADC_ResetCalibration(ADC1);
      while(ADC_GetResetCalibrationStatus(ADC1));
      ADC_StartCalibration(ADC1);
      while(ADC_GetCalibrationStatus(ADC1));
      ADC_SoftwareStartConvCmd(ADC1, ENABLE);
}

void init_USART(){

      RCC_APB2PeriphClockCmd(RCC_APB2Periph_GPIOA |
RCC_APB2Periph_USART1, ENABLE);

      GPIO_InitTypeDef GPIO_InitStructure;
      GPIO_InitStructure.GPIO_Pin = GPIO_Pin_9;
      GPIO_InitStructure.GPIO_Speed = GPIO_Speed_2MHz;
      GPIO_InitStructure.GPIO_Mode = GPIO_Mode_AF_PP;
      GPIO_Init(GPIOA, &GPIO_InitStructure);

      GPIO_InitStructure.GPIO_Pin = GPIO_Pin_10;
      GPIO_InitStructure.GPIO_Mode = GPIO_Mode_IN_FLOATING;
      GPIO_Init(GPIOA, &GPIO_InitStructure);

      USART_InitTypeDef USART_InitStructure;
      USART_InitStructure.USART_BaudRate = 9600;
      USART_InitStructure.USART_WordLength =
USART_WordLength_8b;
      USART_InitStructure.USART_StopBits = USART_StopBits_1;
      USART_InitStructure.USART_Parity = USART_Parity_No;
```

```
       USART_InitStructure.USART_HardwareFlowControl =
USART_HardwareFlowControl_None;
       USART_InitStructure.USART_Mode = USART_Mode_Tx |
USART_Mode_Rx;
       USART_Init(USART1, &USART_InitStructure);
       USART_Cmd(USART1, ENABLE);

}
```

6.6.4. PC code

Here is the C# code:

```csharp
using System;
using System.IO.Ports;
using System.Windows.Forms;

namespace SerialSample6
{
    public partial class Form1 : Form
    {
        SerialPort ComPort = new SerialPort("COM1");
        int val = 0;
        public Form1()
        {
            InitializeComponent();
            if (ComPort.IsOpen) ComPort.Close();
            ComPort.BaudRate = 9600;
            ComPort.Parity = Parity.None;
            ComPort.StopBits = StopBits.One;
            ComPort.DataBits = 8;
            ComPort.Handshake = Handshake.None;
            ComPort.DataReceived += OnSerialDataReceived;
            ComPort.Open();
        }

        private void OnSerialDataReceived(object sender,
SerialDataReceivedEventArgs args)
        {
            while (ComPort.ReadByte() != 0xff);
            char[] c = new char[4];
            ComPort.Read(c, 0, 4);
            string strVal = new string(c);
```

```
        try
        {
            val = Convert.ToInt32(strVal, 16);
        } catch (System.FormatException) {
        }
        ComPort.ReadExisting();
        lblValue.Invoke(new
EventHandler(lblValue_DataReceived));
        }

    private void lblValue_DataReceived(object sender,
EventArgs e)
        {
            lblValue.Text = val.ToString();
            progressBar1.Value = val;
            if (val > 0) progressBar1.Value = val - 1;
            progressBar1.Value = val;
            double valf = (double)val / 399.6;
            lblVoltage.Text = valf.ToString("0.0000") + " V";
            ComPort.Write("x");
        }
    }
}
```

The above example looks similarly to previous one. The only difference is in incoming data handler. First of all we are waiting for delimiting byte, we are reading incoming bytes until we encounter FFh. At this moment we can read four consecutive characters from which the value consists. When we are done with them we can convert into string and then into number. Next we have to do an important thing. Every time an event for incoming data is fired, there are some data waiting in a buffer. We are reading only some of them, 9 bytes at most. When next event occurs we read not new data but previously unread data. This means that we are reading data which are more and more outdated. The effect is that the application doesn't respond immediately to ADC value changes and the lag is constantly growing. The solution is to call *ReadExisting()* function which empties the serial port buffer.

As you can see, the line of code responsible for parsing string value is inside the try-catch clause. Normally it's not necessary but is helpful in when you connect or disconnect serial cable or press microcontroller's reset button. In such cases communication is disturbed and the string variable can contain invalid value which cannot be parsed correctly.

6.7. Displaying received value on 7-seg LED display

In this chapter we are going to tell less about serial transmission but more about peripherals. Let's consider 7-segment LED displays.

6.7.1. Hardware configuration

A simple LED display is used for presenting one or more digits. It consists of LED diodes arranged in a way that any one of ten decimal digits can be made by creating appropriate combination of up to seven segments. Often we can talk about 8-segment displays because there is a decimal dot too. So a LED display is just a set of eight LEDs. Or multiplicity of eight if there is more than one digit. Hence you could expect 16 pins for each digit. In practice however the number of pins is lower and is equal to 10 or less for each digit. This is a result of the fact that dealing with big number of pins or connections is not convenient. Thus a common setup is where all cathodes or all anodes of LEDs are connected together. Then remaining pins can be used for driving particular segments. Using 7-segment (or 8-segment) displays we can present all ten digits and symbols resembling some letters, e.g. A, b, c, C, d, f, h, H, n, P, U, and u.

A similar issue is when we have more than one digit. How many wires would we need for example for four digits? Using standard approach we would end up with 32 connections, which means that a microcontroller with large number of ports is required and that a PCB will get complicated. A solution is to use multiplexing. This approach is used in ZL15AVR development board, look at the schematic as this is a good example. If you don't have ZL15AVR you can download the schematic from Internet.

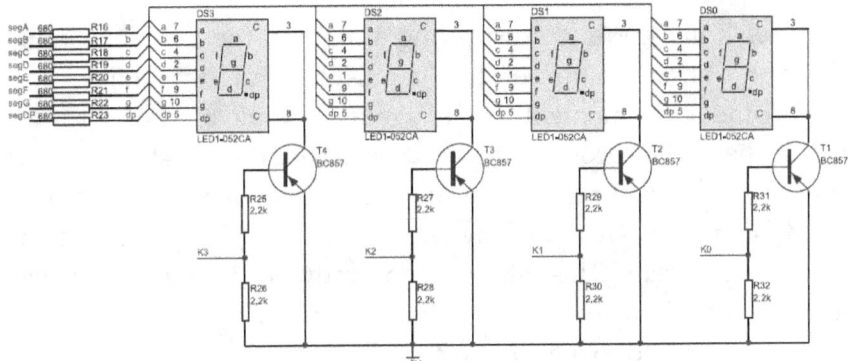

Fig. 6.7.1.1. LED segments on ZL15AVR development board

There are four 1-digit displays. Each one of them has 10 pins: 8 for segments and 2 for common anode. The common anode can be connected to VCC through a transistor by providing low logic level to one of K0..K3 points. Cathodes have serial resistors for limiting current to a safe value. As you can see there are two things which have to be done in order to turn on particular segment of one of these displays. There must be logic 0 at a K0..K3 point and logic 0 at segA..segG or segDP point as well. Simply speaking you have to turn on particular digit and particular segment. In this hardware setup both things are done with low logic level. As you can see, with multiplexing we can control four 8-segment displays using just 12 signals, which means that we need one 8-bit port and half of another one. 12 lines are definitely less than 32 and the circuit is much simpler.

But you can notice a problem: with such an arrangement it's impossible to have different digits on different displays. This is a result of the fact that all segments of the same type are connected together, all 'a' segments are connected together, all 'b' segments are connected together and so on. And yes, you are right; we can't have '1' on first display and '2' on the second one. We can't at the same time. So how the whole thing works? We turn on the first display and provide appropriate combination of zeros and ones to segA..segG lines. We are displaying a digit on the first display. But after a short time we turn off the first display and turn on the second one. At the same moment we change the combination of zeros and ones to have the second digit. And again after some short time we switch to display number three and then to the fourth one. When we are done with fourth display we go back to the first one and everything repeats. This way we constantly go through all displays but only one is active at any given moment. So to have simpler hardware we need to have more complicated software.

The task for a microcontroller is to provide signals for segments and to turn on and off consecutive displays. It must be done fast enough for an eye not to see blinking. A common approach is to have a buffer which is constantly read during and interrupt. A program does its tasks and writes to this buffer when it needs to display something. The interrupt routine constantly reads this buffer and updates displays. This way displaying is separated from the main program and runs quite independently. An opposite solution is also possible: main program is responsible for driving displays and data are updated in interrupts from external sources, e.g. USART. In this chapter however we are going to analyze the first case. It will be also a good chance to say something about timers. If you want you can write multiplexing code not using interrupts, just delays in main loop. The resulting code will be likely not flexible and hard to modify so it's better to try interrupts.

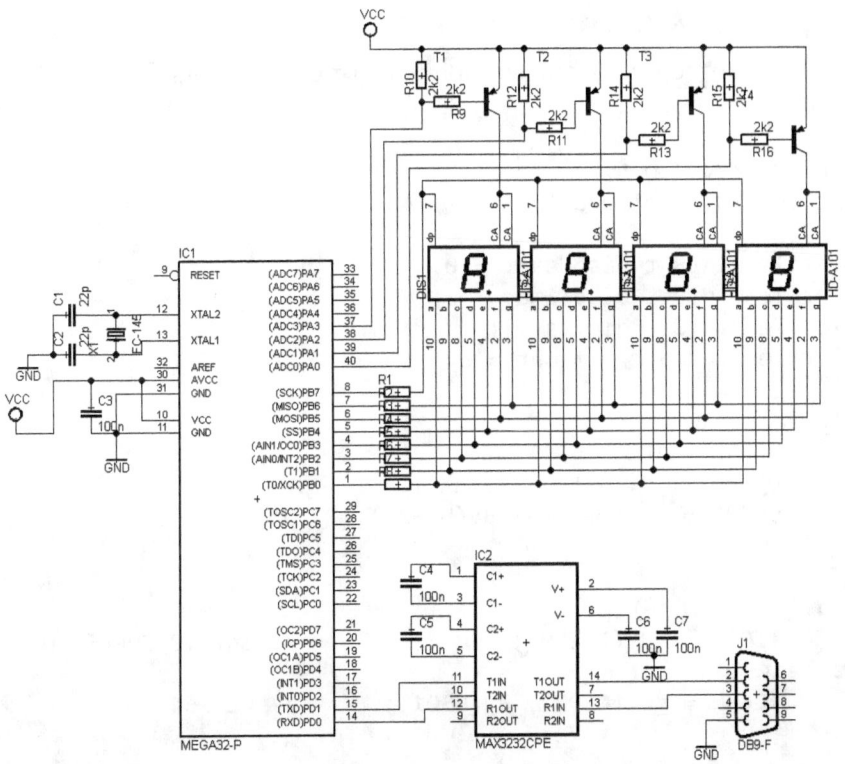

Fig. 6.7.1.2. Circuit with ATmega32.

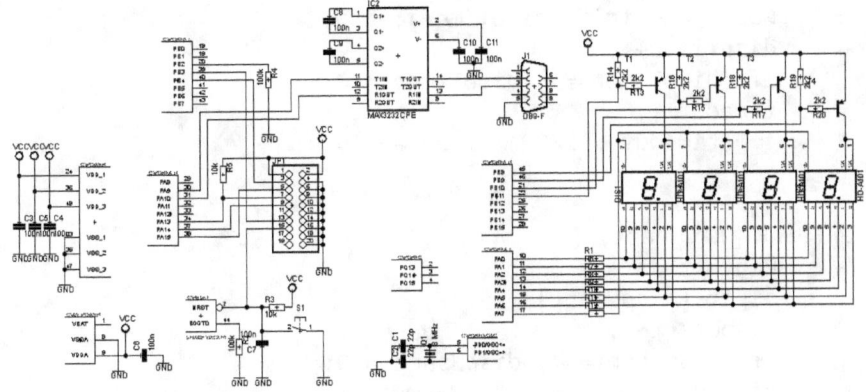

Fig. 6.7.1.3. Circuit with STM32F103CBT6.

281

6.7.2. AVR code

In this example everything happens during interrupts. Let's look at details.

```c
#include <avr/io.h>
#include <avr/interrupt.h>

volatile uint8_t dispUsart = 0;
volatile uint8_t dispTimer = 0;
volatile uint8_t digits[4];
volatile uint16_t timeout = 0;

void init_UART() {
        UBRRL = 103;
        UCSRB = _BV(RXEN)|_BV(RXCIE);
        UCSRC = _BV(URSEL)|_BV(UCSZ1)|_BV(UCSZ0);
}

void init_Timer() {
        TIMSK = _BV(TOIE0);                     // Int T0 Overflow
enabled
        TCCR0 = _BV(CS00)|_BV(CS01);    // CLK/64
}

ISR(TIMER0_OVF_vect)
{
        PORTB = ~_BV(dispTimer);
        PORTA = ~digits[dispTimer];
        dispTimer++;
        if (dispTimer==4) dispTimer = 0;
        timeout++;
        if (timeout > 1000) dispUsart = 0;
}

ISR(USART_RXC_vect) {
        uint8_t data = UDR;
        digits[dispUsart] = data;
        dispUsart++;
        if (dispUsart==4) dispUsart = 0;
        timeout = 0;
}
```

```
int main(void)
{
      DDRA = 0xff;
      DDRB = 0x0f;
      init_UART();
      init_Timer();
      sei();
      while(1) {
      }
}
```

USART initialization is performed as previously, we enable receiver and its interrupt. A new thing is introduction of timer. It's Timer0 which is simple 8-bit timer/counter. By setting the *TOIE0* in initialization routine we enable an interrupt every time the counter register overflows. The timer works in normal mode, it counts from 0 to 255 and then overflows. How quickly the counter is incremented? It depends on bits in *TCCR0* register. With *CS00* and *CS01* bits set the counter is updated every 64 clock ticks. In other words timer gets clocking frequency equal to system frequency divided by 64. For 16 MHz system clock it gives 250 kHz. As the Timer0 is an 8-bit timer with an interrupt generated at overflow we get that frequency divided further by 256 which gives about 976 Hz. This means that an interrupt happens about every millisecond.

Now let's focus on interrupt handler routine. It is responsible for turning on and off appropriate segments of the LED displays. Every time the interrupt happens one digit is served. Index of current digit is stored in the *dispTimer* variable. First we enable current digit using PORTB which is connected to anodes of LED displays. Then we turn on appropriate segments using an array called *digits* which stores combinations of segments for each digit. This array is also indexed by *dispTimer*. The next thing to do is to increment the indexing variable so next time the interrupt happens another display can be updated. As we have only four displays we are zeroing the indexer if it reaches

4. There are two more lines in the interrupt handler but we are going to cover them later.

The interrupt handler for incoming data from serial port is quite simple. It simply writes received value to the *digits* array so it can be displayed when timer interrupt occurs. The array is indexed by *dispUsart* variable which is incremented each time. Again, if array bound is reached the indexer is zeroed.

In this example we are dealing with transmitting multibyte data. Thus there is the problem of communication protocol. For the first sight the situation is simple, we are just transmitting four bytes which are stored in microcontroller's RAM memory and then used to drive appropriate segments of LED displays. But what happens if the transmission is interrupted? If we send four bytes the indexer in microcontroller's code can be set to a value different than 0, therefore digits will be displayed in wrong order. You can check it using PuTTY or similar program. Start the program on the microcontroller and connect to it using PuTTY. Press some key, segments of first display get updated. Press another key, segments of second display get updated. What it you want to change what first display shows? You can't because the third one will get modified. A simple solution used in this example it a timeout. Simply if there is no transmission for some time the indexing variable is set to zero. Hence if new bytes get received displays will be updated starting from the first one. The timeout functionality is realized by a value incremented in Timer0 interrupt handler. If the variable excesses some value, the indexer is reset. In this example Timer0 period is about 1 ms so with the threshold of 1000 the timeout is about one second.

The *main()* function is very simple and contains just initialization code. It configures all pins of port A as outputs because they driver segments of LED displays. Similarly four of port B pins are configured as outputs because they are used to select displays. Then it initializes USART and Timer0 and globally enables interrupts.

6.7.3. STM32 code

The new thing we deal with in this example is a timer. Let's see how it is configured.

```c
#include "stm32f10x.h"
#include <stdio.h>

void init_USART();
void init_Timer();
void init_GPIO();

volatile uint8_t dispUsart = 0;
volatile uint8_t dispTimer = 0;
volatile uint8_t digits[4] = {0,0,0,0};
volatile uint16_t timeout = 0;

int main(void)
{
        init_USART();
        init_GPIO();
        init_Timer();
        while(1);

}

void TIM2_IRQHandler() {
        TIM_ClearFlag(TIM2, TIM_FLAG_Update);
        GPIO_SetBits(GPIOB, 0x0f00);
        GPIO_ResetBits(GPIOB, (1 << (dispTimer+8)));
        GPIO_SetBits(GPIOA, 0x00ff);
        GPIO_ResetBits(GPIOA, digits[dispTimer]);
        dispTimer++;
        if (dispTimer==4) dispTimer = 0;
        timeout++;
        if (timeout > 1000) dispUsart = 0;
}

void USART1_IRQHandler(void)
{
        if (USART_GetITStatus(USART1, USART_IT_RXNE) != RESET)
        {
                uint8_t data = USART_ReceiveData(USART1);
```

```
                digits[dispUsart++] = data;
                if (dispUsart==4) dispUsart = 0;
                timeout = 0;
        }
}

void init_GPIO() {
        RCC_APB2PeriphClockCmd(RCC_APB2Periph_GPIOA |
RCC_APB2Periph_GPIOB, ENABLE);

        GPIO_InitTypeDef GPIO_InitStructure;
        GPIO_InitStructure.GPIO_Pin = GPIO_Pin_0 | GPIO_Pin_1 |
GPIO_Pin_2 | GPIO_Pin_3 | GPIO_Pin_4 | GPIO_Pin_5 |
GPIO_Pin_6 | GPIO_Pin_7;
        GPIO_InitStructure.GPIO_Speed = GPIO_Speed_2MHz;
        GPIO_InitStructure.GPIO_Mode = GPIO_Mode_Out_OD;
        GPIO_Init(GPIOA, &GPIO_InitStructure);

        GPIO_InitStructure.GPIO_Pin = GPIO_Pin_8 | GPIO_Pin_9 |
GPIO_Pin_10 | GPIO_Pin_11;
        GPIO_Init(GPIOB, &GPIO_InitStructure);
}

void init_USART(){

        RCC_APB2PeriphClockCmd(RCC_APB2Periph_GPIOA |
RCC_APB2Periph_USART1, ENABLE);

        GPIO_InitTypeDef GPIO_InitStructure;
        GPIO_InitStructure.GPIO_Pin = GPIO_Pin_9;
        GPIO_InitStructure.GPIO_Speed = GPIO_Speed_2MHz;
        GPIO_InitStructure.GPIO_Mode = GPIO_Mode_AF_PP;
        GPIO_Init(GPIOA, &GPIO_InitStructure);

        GPIO_InitStructure.GPIO_Pin = GPIO_Pin_10;
        GPIO_InitStructure.GPIO_Mode = GPIO_Mode_IN_FLOATING;
        GPIO_Init(GPIOA, &GPIO_InitStructure);

        USART_InitTypeDef USART_InitStructure;
        USART_InitStructure.USART_BaudRate = 9600;
        USART_InitStructure.USART_WordLength =
USART_WordLength_8b;
        USART_InitStructure.USART_StopBits = USART_StopBits_1;
```

```
        USART_InitStructure.USART_Parity = USART_Parity_No;
        USART_InitStructure.USART_HardwareFlowControl =
USART_HardwareFlowControl_None;
        USART_InitStructure.USART_Mode = USART_Mode_Tx |
USART_Mode_Rx;
        USART_Init(USART1, &USART_InitStructure);
        USART_ITConfig(USART1, USART_IT_RXNE, ENABLE);
        USART_Cmd(USART1, ENABLE);

        NVIC_PriorityGroupConfig(NVIC_PriorityGroup_0);
        NVIC_InitTypeDef NVIC_InitStructure;
        NVIC_InitStructure.NVIC_IRQChannel = USART1_IRQn;
        NVIC_InitStructure.NVIC_IRQChannelSubPriority = 0;
        NVIC_InitStructure.NVIC_IRQChannelCmd = ENABLE;
        NVIC_Init(&NVIC_InitStructure);

}

void init_Timer(){

        RCC_APB1PeriphClockCmd(RCC_APB1Periph_TIM2, ENABLE);

        TIM_TimeBaseInitTypeDef TIM_InitStruct;
        TIM_InitStruct.TIM_ClockDivision = TIM_CKD_DIV1;
        TIM_InitStruct.TIM_CounterMode = TIM_CounterMode_Up;
        TIM_InitStruct.TIM_Period = 255;
        TIM_InitStruct.TIM_Prescaler = 64;
        TIM_TimeBaseInit(TIM2, &TIM_InitStruct);

        TIM_ClearFlag(TIM2, TIM_FLAG_Update);
        TIM_ITConfig(TIM2, TIM_IT_Update, ENABLE);
        TIM_Cmd(TIM2, ENABLE);

        NVIC_InitTypeDef NVIC_InitStruct;
        NVIC_InitStruct.NVIC_IRQChannel = TIM2_IRQn;
        NVIC_InitStruct.NVIC_IRQChannelCmd = ENABLE;
        NVIC_InitStruct.NVIC_IRQChannelPreemptionPriority = 0;
        NVIC_InitStruct.NVIC_IRQChannelSubPriority = 0;
        NVIC_Init(&NVIC_InitStruct);
}
```

The *init_Timer()* function is responsible for configuring a timer and its interrupts. Here we are using TIM2 timer. We set the *TIM_ClockDivision* field to *TIM_CKD_DIV1* but it could be *TIM_CKD_DIV2* or *TIM_CKD_DIV4* as well. This field indicates the division ratio between the timer clock (CK_INT) frequency and sampling clock used by the digital filters (ETR, TIx) and is not important in our case. The timer's counter is configured to count up. The *TIM_Period* field defines maximum value of the counter. Here the value is 255 which means, that the timer counts from 0 to 255 and then starts again from 0. This gives us 256 states. The timer has also a prescaler which introduces additional clock division. Here it is 64. If the main clock is 8 MHz then the counter is incremented with the frequency equal to 125 kHz. The overflow interrupt is hence generated about 490 times per second. We don't need any specific frequency here. It just can't be too low because then blinking of LED displays starts to be visible. When the timer is configured we continue with initialization. The overflow event is called update by STM. We want to clear the overflow flag so we call *TIM_ClearFlag* with *TIM_FLAG_Update* parameter. Then we enable the overflow (update) interrupt and enable the timer itself. Finally we enable timer interrupt with NVIC.

LED displays are driven in a standard, multiplexed way. In any given moment only one display is active. Number of current display is stored in the *dispTimer* variable and is activated by setting appropriate bits of port B. Segments of LCD displays are driven by pins of port A. When appropriate bits are set the number of current display is incremented for next interrupt. If the last display is reached then the *dispTimer* variable is set back to zero. Data for each display are taken from the *digits* buffer which is filled by USART1 interrupt routine. This function uses separate variable for indexing displays: *dispUsart*. This variable is set to zero by the timer interrupt if no data are received for more than about 2 seconds. This is a simple method of avoiding a situation where due to disconnected cable or some other problem not all four bytes were received. Without the variable automatically set back to zero there would be a problem when the transmission is resumed because there would be a shift of displayed data, e.g. data for display 1 shown on display 3, data for display 2 shown on display 4 and so on.

A few words about port B pins selection. Pins 8 – 11 are have been chosen because pins like 3 and 4 are used by JTAG and pin 2 serves also as BOOT1. Hence pins 8 – 11 were selected.

One more thing. When we set pins we do it like this:

```
GPIO_SetBits(GPIOB, 0x0f00);
GPIO_SetBits(GPIOA, 0x00ff);
```

This sets all four pins of port B we use and all eight pins of port A which we use. In the following lines we reset some pins. Thus some pins are set and then immediately reset. It's not a problem but if you want to set only those pins which should be set when you can do it like this:

```
GPIO_SetBits(GPIOB, 0x0f00 & ~(1 << (dispTimer+8)));
GPIO_SetBits(GPIOA, 0x00ff & ~digits[dispTimer]);
```

6.7.4. PC code

This example doesn't introduce much novelty in terms of data transmission. The program just sends contents of 4-byte buffer. The interesting thing is how data provided by user are prepared to be sent. Here is the code:

```
using System;
using System.IO.Ports;
using System.Windows.Forms;

namespace SerialSample7
```

```
{
    public partial class Form1 : Form
    {
        SerialPort ComPort = new SerialPort("COM1");

        byte[] segmentSets = new byte[]{0x3F, 0x06, 0x5B,
0x4F, 0x66, 0x6D, 0x7D, 0x07, 0x7F, 0x6F };
        public Form1()
        {
            InitializeComponent();
            if (ComPort.IsOpen) ComPort.Close();
            ComPort.BaudRate = 9600;
            ComPort.Parity = Parity.None;
            ComPort.StopBits = StopBits.One;
            ComPort.DataBits = 8;
            ComPort.Handshake = Handshake.None;
            ComPort.Open();
        }

        private void btnSet_Click(object sender, EventArgs e)
        {
            byte[] digits = new byte[4];
            stringToSegments(txtDigits.Text, digits);
            ComPort.Write(digits, 0, 4);
        }

        private void stringToSegments(String str, byte[]
digits)
        {
            bool dot = false;
            int j = 0;
            for (int i = 0; i < 4; i++) digits[i] = 0;
            for (int i = str.Length - 1; i >= 0; i--)
            {
                char c = str[i];
                if (c == '.' && dot)
                {
                    digits[j] = 0x80;
                    j++;
                    continue;
                }
                if (c == '.') dot = true;
                if (Char.IsDigit(c))
```

```
            {
                digits[j] = segmentSets[c - '0'];
                if (dot) digits[j] += 0x80;
                dot = false;
                j++;
                continue;
            }
        }
        if (dot) digits[j] = 0x80;
    }

    private void timer1_Tick(object sender, EventArgs e)
    {
        byte[] digits = new byte[4];
        string dot = "";
        if (DateTime.Now.Millisecond > 500) dot = ".";
        string time = DateTime.Now.Hour.ToString("00") +
dot + DateTime.Now.Minute.ToString("00");
        stringToSegments(time, digits);
        ComPort.Write(digits, 0, 4);

        if (chkUpdate.Checked)
        {
            txtDigits.Text = time;
            string timeHex = digits[3].ToString("X2") +
digits[2].ToString("X2") + digits[1].ToString("X2") +
digits[0].ToString("X2");
            maskedTextBox1.Text = timeHex;
        }
    }

    private void chkTime_CheckedChanged(object sender,
EventArgs e)
    {
        timer1.Enabled = chkTime.Checked;
    }

    private void btnHex_Click(object sender, EventArgs e)
    {
        byte[] digits = new byte[4];
        string[] digitsStr =
maskedTextBox1.Text.Split(new char[] {' '});
```

```
          for (int i = 0; i < 4; i++) digits[3 - i] =
Convert.ToByte(digitsStr[i], 16);
          ComPort.Write(digits, 0, 4);
      }
   }
}
```

This application has three functions: sending string of digits with optional dots, sending arbitrary values and sending current time. These are examples of how data for displays can be provided.

The first functions assumes that a user enters up to four digits and those digits are displayed on LED displays. The additional assumption is that also dots on the displays can be turned on. This leads to following convention. Digits are displayed as they are. If there is a dot after a digit then the dot is displayed on the same display as the digit. If a dot is placed after another dot or is a first character in the string than it is displayed on its own. This is simple algorithm for easy and quick entering numbers and dots to be displayed. It doesn't allow e.g. to display a digit on display 1 and a dot on display 0. But it's just an example and you can improve it if you want.

An important part on the code is a parser which implements above convention and allows to provide values of bytes to send basing on entered string. The *stringToSegments()* function is quite simple. It goes from the end of string to the beginning and checks if current character is a digit or dot. If it is a dot it remembers the fact. If it is a digit it looks for appropriate value from the *segmentSets* array which stores values representing combinations of segments for each decimal digit. If there was a digit before, the value 0x80 is added so the dot segment is turned on as well. If there is a dot and there was a dot previously then only a dot is displayed. If a dot is the last character (looking from the right hand side) that it is also displayed solely.

Generally this is just an example. You can write different parser or come up with different convention of strings, for example using not only digits and dots but also spaces. Or you may decide that you need just digits, then the parser would be very simple. Anyway, everything is up to you.

The parser described above is used in two places. The first one is when the *btnSet* button is clicked. There it prepares array of bytes to be sent basing on text entered into *txtDigits*. The second one is when the LED displays are used for a digital clock. In this scenario the PC application constantly sends current time to the microcontroller and it displays it. This is realized using a timer which fires every 100 milliseconds. When it happens, current time is obtained and stored in a string variable. Then this string is parsed to get values ready to be sent to the microcontroller. An additional feature is a blinking dot which is displayed for half of each second.

The last function allows you to turn on and off arbitrary segments of all displays. You can enter four hexadecimal values representing segments in the displays. For entering values a masked text box control was used. Entered numbers are parsed and sent to the microcontroller.

When running in the digital clock mode there can be enabled a function for updating both text boxes. This way you can see what is actually sent to the microcontroller. You can turn it off if you have some values entered and don't want them to be overwritten in clock mode.

Remark

In this example the logic of converting numbers into combinations of segments is placed in the PC application and the microcontroller has just basic code for driving LEDs in displays. A different approach is also possible, where numbers or some more complicated string is sent to a microcontroller and its chip's task to analyze the data and turn on/off appropriate segments. It depends on particular case. It's a matter of where the logic for processing data is placed, in the PC application or in the microcontroller's code. It's not a matter of driving LED displays so we are just hinting about the subject.

6.8.   *Displaying received text on an LCD display*

In the previous chapter we were dealing with standard LED displays. They are good for digits and simple symbols but not for text. The most popular text mode LCD displays are those based on HD44780 driver or compatible chips. From those LCDs the 2x16 models are most common, they can show two lines of text, each up to 16 characters long. Alphanumeric LCD displays are both easier and harder to control. Easier, because they have built-in set of characters so we don't have to build letters from individual pixels. Harder because they have their protocol of communication which we need to follow. This protocol is not complicated however and writing own code for controlling alphanumeric LCDs is not a hard task. There are also ready to use libraries available on the Internet.

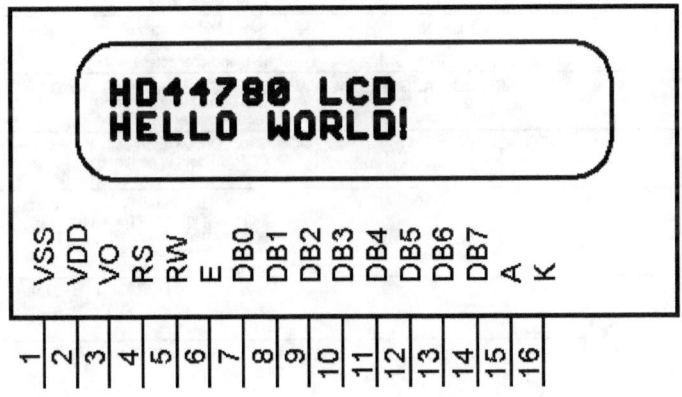

Fig. 6.8.1. Pinout of typical 2x16 LCD compatible with HD44780

Communication with an alphanumeric LCD is done using data lines and control lines. The display has 8 data lines. You can communicate with the LCD in 8-bit mode and use all those lines or in 4-bit mode and use just half of them. As you can guess the 8-bit mode is faster because whole byte is being sent at once. In 4-bit mode each byte is sent in two steps, first 4 more significant bits, then 4 less significant bits. The advantage of 4-bit mode is simpler circuit and possibility of using a microcontroller with low number of free I/O pins. Moreover difference in speed is negligible.

Instruction	RS	R/W	DB7	DB6	DB5	DB4	DB3	DB2	DB1	DB0	Description	Execution Time (max) (when f_{cp} or f_{OSC} is 270 kHz)
Clear display	0	0	0	0	0	0	0	0	0	1	Clears entire display and sets DDRAM address 0 in address counter.	
Return home	0	0	0	0	0	0	0	0	1	—	Sets DDRAM address 0 in address counter. Also returns display from being shifted to original position. DDRAM contents remain unchanged.	1.52 ms
Entry mode set	0	0	0	0	0	0	0	1	I/D	S	Sets cursor move direction and specifies display shift. These operations are performed during data write and read.	37 µs
Display on/off control	0	0	0	0	0	0	1	D	C	B	Sets entire display (D) on/off, cursor on/off (C), and blinking of cursor position character (B).	37 µs
Cursor or display shift	0	0	0	0	0	1	S/C	R/L	—	—	Moves cursor and shifts display without changing DDRAM contents.	37 µs
Function set	0	0	0	0	1	DL	N	F	—	—	Sets interface data length (DL), number of display lines (N), and character font (F).	37 µs
Set CGRAM address	0	0	0	1	ACG	ACG	ACG	ACG	ACG	ACG	Sets CGRAM address. CGRAM data is sent and received after this setting.	37 µs
Set DDRAM address	0	0	1	ADD	ADD	ADD	ADD	ADD	ADD	ADD	Sets DDRAM address. DDRAM data is sent and received after this setting.	37 µs
Read busy flag & address	0	1	BF	AC	AC	AC	AC	AC	AC	AC	Reads busy flag (BF) indicating internal operation is being performed and reads address counter contents.	0 µs
Write data to CG or DDRAM	1	0	Write data								Writes data into DDRAM or CGRAM.	37 µs $t_{ADD} = 4$ µs*
Read data from CG or DDRAM	1	1	Read data								Reads data from DDRAM or CGRAM.	37 µs $t_{ADD} = 4$ µs*

I/D = 1:	Increment	DDRAM:	Display data RAM	Execution time
I/D = 0:	Decrement	CGRAM:	Character generator	changes when
S = 1:	Accompanies display shift		RAM	frequency changes
S/C = 1:	Display shift	ACG:	CGRAM address	Example:
S/C = 0:	Cursor move	ADD:	DDRAM address	When f_{cp} or f_{osc} is
R/L = 1:	Shift to the right		(corresponds to cursor	250 kHz,
R/L = 0:	Shift to the left		address)	$37 \text{ µs} \times \frac{270}{250} = 40 \text{ µs}$
DL = 1:	8 bits, DL = 0: 4 bits	AC:	Address counter used for	
N = 1:	2 lines, N = 0: 1 line		both DD and CGRAM	
F = 1:	5×10 dots, F = 0: 5×8 dots		addresses	
BF = 1:	Internally operating			
BF = 0:	Instructions acceptable			

Note: — indicates no effect.

Fig. 6.8.2. LCD instruction set

There are three control lines: RS, R/W and EN. In write mode the RS line tells the LCD whether we are going to send a character to be displayed or a command. If RS is low (logic zero) then LCD treats a byte present on data lines as a command to be executed. There are commands for such operations like turning the LCD on and off, clearing display, enabling/disabling cursor, moving cursor, enabling/disabling blinking of the cursor, shifting text or defining custom characters. If RS is high (logic one) then the byte supplied to data lines is interpreted as ASCII code of a character to be displayed or as definition of a custom character. Depending on selected address, data supplied by user go to DD RAM which contains characters to be displayed, or to CG RAM which stores definitions of user defined characters. In read mode the RS line is used to select between reading DD RAM (or CG RAM) address and reading DD RAM (CG RAM) data.

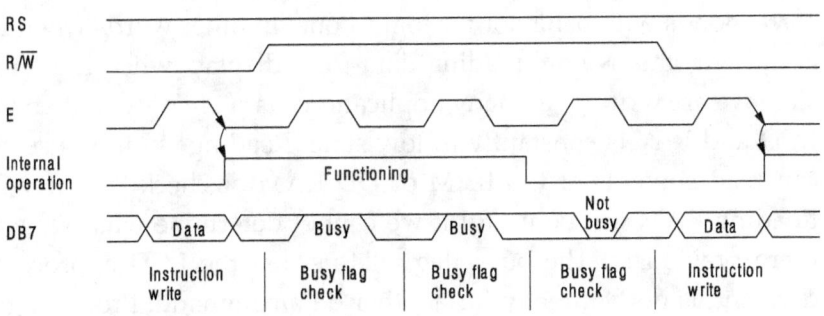

Fig. 6.8.3. Communication in 8-bit mode

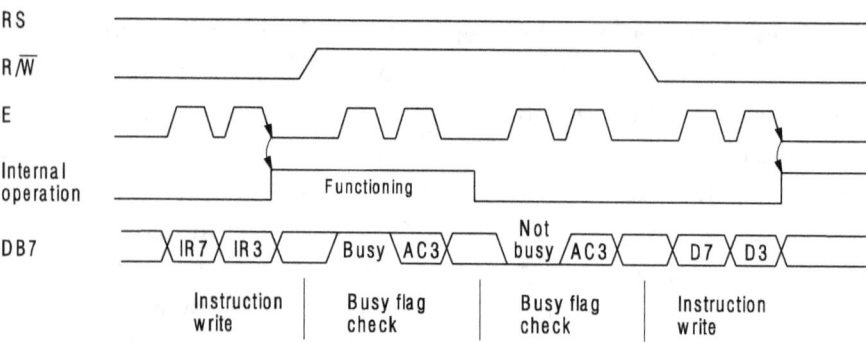

Note: IR7 , IR3 are the 7th and 3rd bits of the instruction.
AC3 is the 3rd bit of the address counter.

Fig. 6.8.4. Communication in 4-bit mode

The R/W line sets data direction. As the "/" characters means negation, we can read the name of the line as "read, not write" which corresponds with behavior for logical one. In other words when R/W is in high state we are reading data from display, when it is in low state we are writing. In many applications data are only written to an LCD and R/W is constantly in low state. Reading is rarely used, we can read contents of CG RAM or DD RAM or check the busy flag. Reading RAM can be useful if we cannot determine what we wrote there previously. The busy flag tells us that the LCD is processing data and is not yet ready for another data/command. Processing can take some time, from about one hundred microseconds to a few milliseconds. It's good to check this flag because it allows fast and reliable communication. The code below makes use of busy flag checking.

As you see we can read addresses or data. What about the busy flag? You just read current address. The address itself is stored in bits 6..0, bit 7 is the busy flag.

There is also the EN line, sometimes called just E. The name stands for enable. If you wondered when exactly data are written to an LCD here you have an answer: on falling edge on EN. Normally this line is in low state. When we want to send data or command we set proper states of RS, R/W and data lines. When they are ready we change EN to one and after short moment to zero again. Now data are sent to the LCD. What has left is to wait for a short time after LCD finishes processing received data/command.

LCD has small internal read only memory storing information about how every character looks like, it is called CG ROM. There is also a small piece of RAM (called CG RAM) where you can put own characters by defining them pixel by pixel. This memory has 64 bytes and can store 8 custom characters because one character needs 8 bytes. To display those characters just provide their addresses as ASCII codes. Generally the map of available characters looks like this:

- from 0 to 7 – custom characters from CG RAM

- from 16 to 127 – standard ASCII characters

- from 127 to 255 – characters from various alphabets, usually from Far East

To define own characters you have to send a command which contains address of CG RAM cell. From this moment data will go to this and consecutive addresses. When the address is set, send bytes describing pixels in each horizontal line of a character, top to bottom. A byte contains information about pixels of a line in its 5 less significant bits, zeros for light pixels, ones for dark pixels. You can easily calculate values for your character by yourself. There is also an online tool available which allows obtaining values just by clicking appropriate pixels. You can find it at http://www.quinapalus.com/hd44780udg.html When you are done with all lines, you can send another character. If you want to display custom defined character you need to remember about switching from CG RAM to DD RAM. Then you can tell the display to show your character. Let's say we want to define the euro € sign and have it in the first cell of CG RAM. Then we want to display it on the first position of the first line of LCD. The steps are as follows:

- select first cell in CG RAM by sending 0x40 command

- send data bytes describing pixels, for € they are: 0x03, 0x04, 0x1e, 0x08, 0x1e, 0x08, 0x07, 0x00

- select first cell in DD RAM by sending 0x80 command

- display defined character by sending 0x00 as data

We need to tell a little bit about DD RAM. This area of memory stores characters to be shown on display. It has 80 bytes, 40 bytes for each line (for 2x16 LCD). This means that in fact we have logically 2x40 display but only 16 characters from each line are visible at any given time. Does this mean that 2x24 bytes are wasted? Not exactly. The display can be scrolled left or right and characters which were in an invisible region can therefore be made visible. This allows not only for creating fancy effects but also to change displayed text just by scrolling. Of course the new text must be written to the invisible region before.

Both lines are looped; when the text is scrolled left it will start to appear from the right and vice versa. But because the visible area is just part of whole line will be not visible immediately, but after the invisible part is scrolled too.

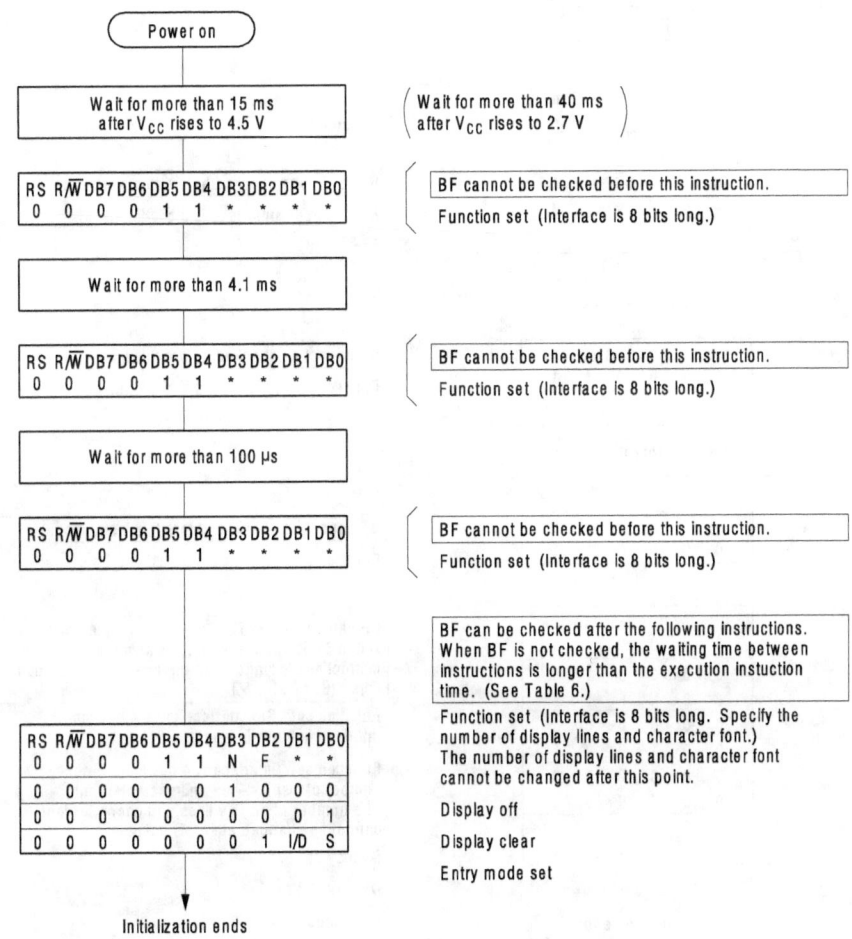

Fig. 6.8.5. Initialization of an LCD in 8-bit mode

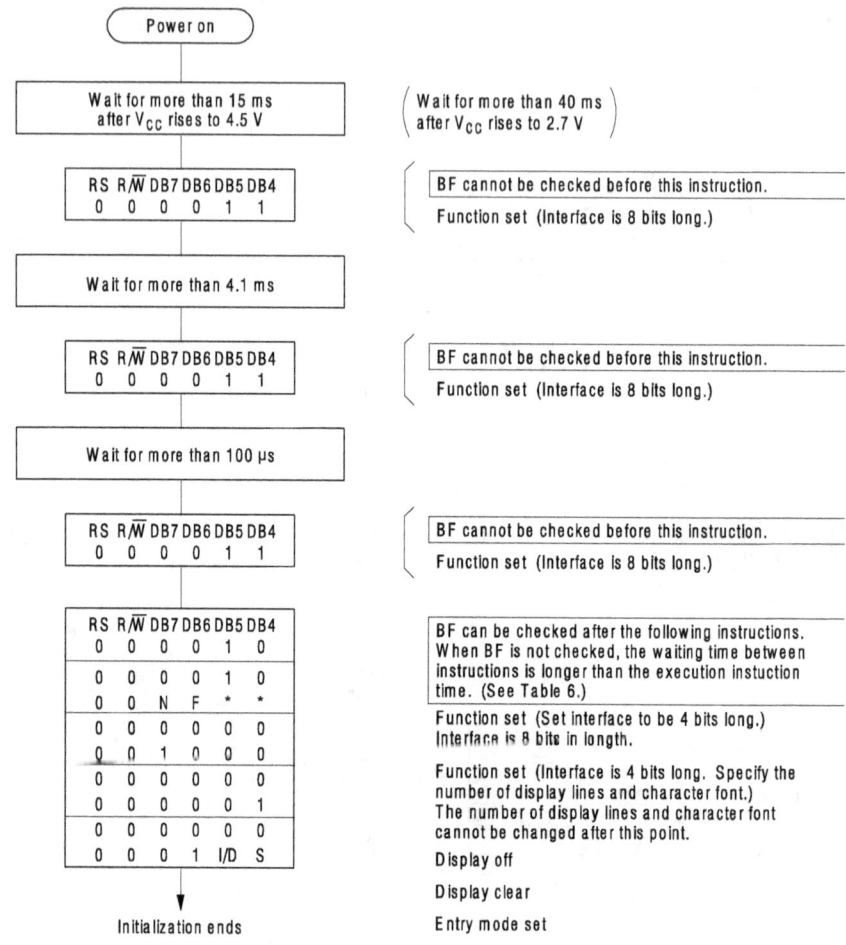

Fig. 6.8.6. Initialization of an LCD in 4-bit mode

So we can say that both lines work independently, we can treat them as two separate circular buffers. But from the point of view of DD RAM they are one thing. When you start writing data to DD RAM you will observe characters appearing on LCD, starting from upper left corner. When 16 characters are transmitted, further ones will be placed in the invisible part. But what happens when you write 41st character? It will show up on the beginning of second line. So from the point of view of DD RAM we deal with one 80-characters long line.

6.8.1. Hardware configuration

First we need to decide whether we choose 8-bit or 4-bit communication. For 8-bit everything is simple, just connect data lines (D7..D0) to appropriate pins of some port, e.g. PORTA7..PORTA0. For 4-bit communication connect 4 more significant data pins of LCD (D7..D4) to 4 more significant pins of a port, e.g. PORTA7...PORTA4. RS, R/W, and EN lines should be connected to some free I/O pins.

Of course the display must be powered so don't forget about connecting VCC (pin 2) and GND (pin 1). Usually LCDs have also a backlight. If you want to use it connect pin 15 (anode) to VCC and pin 16 (cathode) do GND. If you wish to control the backlight from your program you can use a transistor. If it is a NPN type, connect collector to cathode and emitter to GND. In case of PNP connect collector to anode and emitter to VCC. In both cases the base should be connected to available microcontroller's port pin through a resistor. Before you choose particular transistor check current drawn by the backlight, it can be pretty large, even more than 200 milliamps. Thus you need to makes sure your transistor can work with such current. As for the transistor the resistance of a few kilohms should be OK. When you write code for the microcontroller remember that the backlight is turned on with NPN transistor for logic one and for logic zero with PNP type.

There is one more thing: contrast. It must be correctly adjusted for optimal visibility. You can use a potentiometer connected to GND, VCC and pin 3 of LCD. A 10 kOhm potentiometer will be OK.

Serial port and microcontrollers

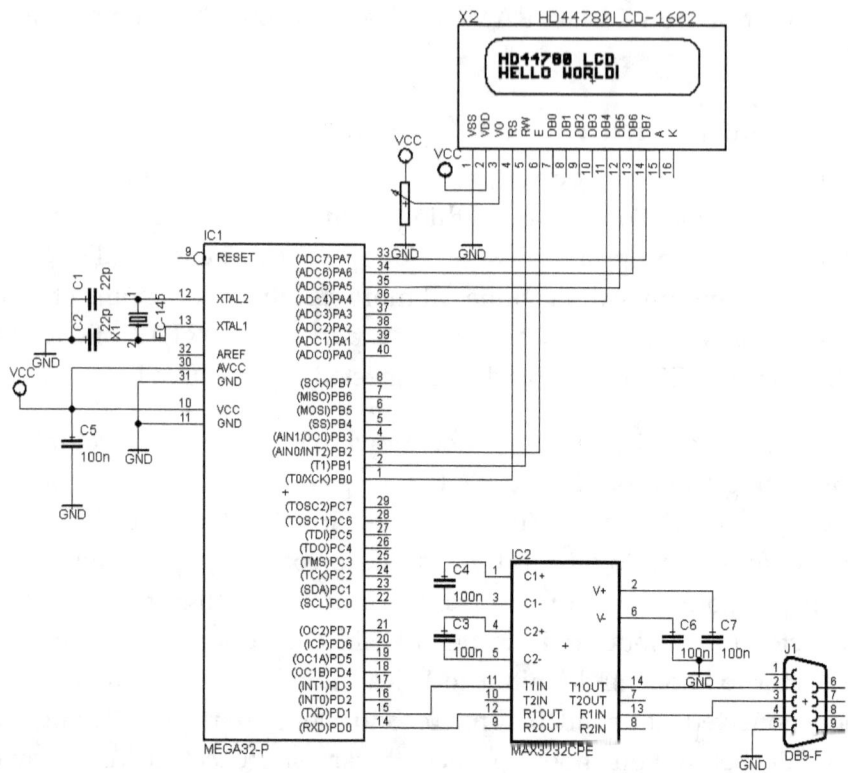

Fig. 6.8.1.1. Circuit with ATmega32.

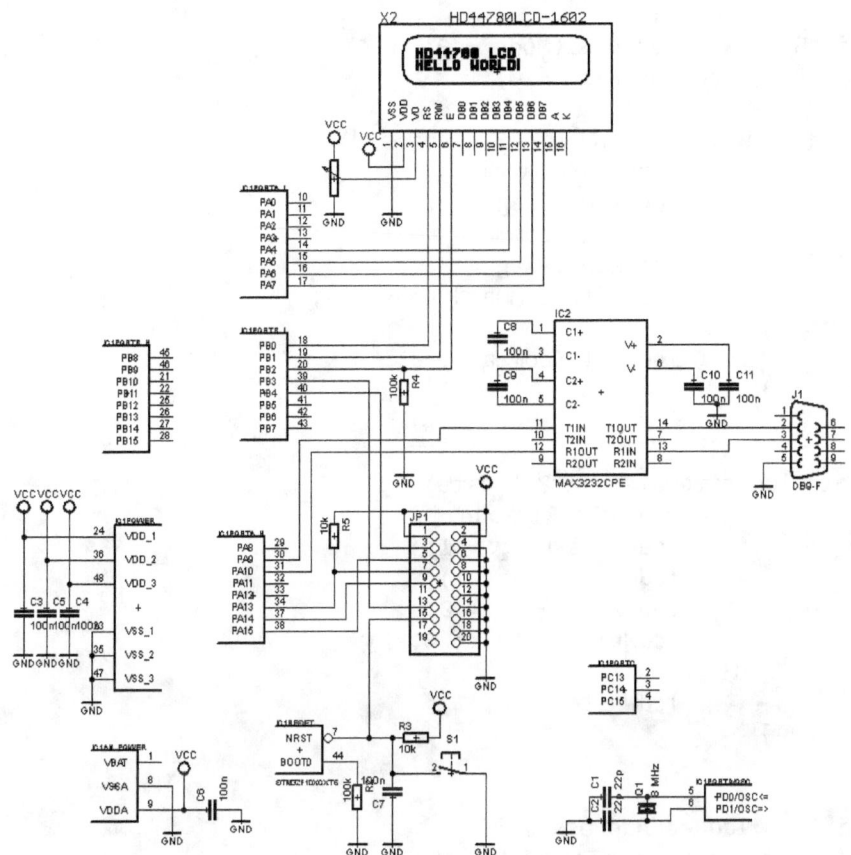

Fig. 6.8.1.2. Circuit with STM32F103CBT6.

6.8.2. AVR code

This time the project is more complicated because it consists of two files: main program and simple LCD library. Let's start with the library. It consists of implementation file *lcd.c* and header file *lcd.h*. The first one looks as follows:

```
#include "lcd.h"
#define F_CPU 16000000UL
#include <util/delay.h>
```

```
uint8_t switchedTo4BitMode = 0;
uint8_t checkBusyFlag = 0;

void lcdWriteCommand(uint8_t command) {
      if (switchedTo4BitMode){
            checkBusyFlag = 1;
            lcdWrite(command, 0);
            checkBusyFlag = 0;
            lcdWrite(command << 4, 0);
      }else {
            lcdWrite(command, 0);
      }
}

void lcdWriteData(uint8_t data) {
      if (switchedTo4BitMode){
            checkBusyFlag = 1;
            lcdWrite(data, 1);
            checkBusyFlag = 0;
            lcdWrite(data << 4, 1);
      }else {
            lcdWrite(data, 1);
      }
}

uint8_t lcdReadAddress() {
      if (switchedTo4BitMode){
            return ((lcdRead(0) & 0xf0) | (lcdRead(0) >>
4));
      }else{
            return lcdRead(0);
      }
}

uint8_t lcdReadData(){
      if (switchedTo4BitMode){
            return ((lcdRead(1) & 0xf0) | (lcdRead(1) >>
4));
      }else{
            return lcdRead(1);
      }
}
```

```
void lcdWrite(uint8_t command, uint8_t rs) {
        if (checkBusyFlag) while (lcdReadAddress() & 0x80)
_delay_us(10);

        // write mode, R/W low
        LCD_CONTROL_PORT &= ~_BV(LCD_RW);
#ifdef BITS4
        LCD_DATA_DDR |= 0xf0; //four upper pins as outputs
#else
        LCD_DATA_DDR = 0xff; //all data pins as outputs
#endif
        //R/S
        if (rs) LCD_CONTROL_PORT |= _BV(LCD_RS); else PORTB &=
~_BV(LCD_RS);
        //command
#ifdef BITS4
        LCD_DATA_PORT &= command | 0xf0; //reset bits
        LCD_DATA_PORT |= command & 0x0f; //set bits
#else
        LCD_DATA_PORT = command;
#endif

        //EN high
        LCD_CONTROL_PORT |= _BV(LCD_EN);
        //wait 1 us
        _delay_us(1);
        //EN low
        LCD_CONTROL_PORT &= ~_BV(LCD_EN);
        //wait 1 us
        _delay_us(1);
}

uint8_t lcdRead(uint8_t rs) {
#ifdef BITS4
        LCD_DATA_DDR &= 0x0f; //four upper pins as inputs
#else
        LCD_DATA_DDR = 0x00; //all data pins as inputs
#endif
        //R/S
        if (rs) LCD_CONTROL_PORT |= _BV(LCD_RS); else
LCD_CONTROL_PORT &= ~_BV(LCD_RS);
        //R/W high
```

```
        LCD_CONTROL_PORT |= _BV(LCD_RW);
        //EN high
        LCD_CONTROL_PORT |= _BV(LCD_EN);
        //wait 1 us
        _delay_us(1);
        //read
        uint8_t data = LCD_DATA_INPUT;
        //EN low
        LCD_CONTROL_PORT &= ~_BV(LCD_EN);
        //wait 1 us
        _delay_us(1);
        return data;
}

void lcdInit() {
        LCD_CONTROL_DDR |= _BV(LCD_EN)|_BV(LCD_RW)|_BV(LCD_RS);
        _delay_ms(15);
        lcdWriteCommand(LCD_COMMAND_FUNCTION_SET |
LCD_OPTION_8BITS); // set interface - 8-bit
        _delay_ms(4);
        lcdWriteCommand(LCD_COMMAND_FUNCTION_SET |
LCD_OPTION_8BITS); // set interface - 8-bit
        _delay_ms(1);
        lcdWriteCommand(LCD_COMMAND_FUNCTION_SET |
LCD_OPTION_8BITS); // set interface - 8-bit
#ifdef BITS4
        _delay_ms(1);
        lcdWriteCommand(LCD_COMMAND_FUNCTION_SET); // set
interface - 4-bit
        _delay_ms(1);
        switchedTo4BitMode = 1;
        lcdWriteCommand(LCD_COMMAND_FUNCTION_SET |
LCD_OPTION_2LINES); // set interface - 4-bit, 2 lines, 5x8
font
#else
        checkBusyFlag = 1;
        lcdWriteCommand(LCD_COMMAND_FUNCTION_SET |
LCD_OPTION_8BITS | LCD_OPTION_2LINES); // set interface - 8-
bit, 2 lines, 5x8 font
#endif
        lcdWriteCommand(LCD_COMMAND_ON); // off
        lcdWriteCommand(LCD_COMMAND_ON | LCD_OPTION_DISPLAY);
// on
```

```
    lcdWriteCommand(LCD_COMMAND_ENTRY_MODE |
LCD_OPTION_INCREMENT); // scroll
    lcdWriteCommand(LCD_COMMAND_CLEAR); // clear
}

void lcdString(char * buf) {
    while (*buf != 0) {
        lcdWriteData(*buf);
        buf++;
    }
}
```

One of the more important functions is *lcdWrite()* which sends one byte (or half of byte in 4-bit mode) to an LCD. First it checks if the display is ready by reading the busy flag. If it is not then after short delay the flag is checked again. When the LCD is ready the R/W line is set to low state because we are going to write data. Then the data pins are configured as outputs. Then, depending whether we are sending a command or data, RS line is set accordingly. The next thing is to set actual data byte on port pins which are connected to LCD data lines. In case of 4-bit communication the situation is a little bit tricky because we don't want to change state of unused pins. Therefore we set states of pins in two steps, each time taking into account the rest of port's pins. To set appropriate pins to zero we use logical AND operation performed on the port. In order not to modify less significant bits we prepare data by performing logical OR on the byte to be written with 0Fh mask. This way we set all 4 less significant bits to 1 which guarantees that during AND operation less significant port's bits won't be modified. For setting port's bits to one the approach is similar. With AND operation and F0h mask we are zeroing 4 less-significant bits. This way port's less significant bits won't be affected by the OR operation. When we are done with setting port bits we are ready to set EN high and low after short moment so the data are received by LCD. After a short delay the function finishes.

The *lcdRead()* function works in analogous way. Data pins are configured as inputs, R/W is set high and RS is set depending on whether we are going to read address (and the busy flag) or data. EN is set high to request data. After short delay data are read. Then EN goes low again and data are returned.

For sending data and commands there are two simple methods: *lcdWriteData()* and *lcdWriteCommand()* which call *lcdWrite()* with appropriate value of the RS line. In 4-bit mode they are responsible for dividing communication into two steps, first more significant half-byte is read/written, then the less significant. Busy flag is checked only before the first write transmission, there is no sense in checking it before the second half-byte being written.

A few words about writing and reading half-bytes. In case of write, during first operation we pass the data byte as it is. In 4-bit mode the *lcdWrite()* function ignores less significant bits so there is no problem. During the second write operation we shift the byte by 4 positions left so the less significant half-byte becomes the more significant and can be sent to an LCD. Reading is a little bit more tricky because *lcdRead()* fetches state of a whole port, even in 4-bit mode. Thus we use logical AND with F0h mask to remove the less significant bits. During the second read operation we want to get less significant bits. Because we use more significant pins of port to communicate with LCD those bits will be places in more significant places. That's why we have to shift them right by four positions. When we have two half-bytes we join them together using logical OR.

The *lcdInit()* function sets control pins as outputs and performs standard initialization routine for 8-bit or 4-bit communication.

Let's take a look at the header file of our LCD library.

```c
#include <avr/io.h>

void lcdWrite(uint8_t command, uint8_t rs);
void lcdWriteCommand(uint8_t command);
void lcdWriteData(uint8_t data);
uint8_t lcdRead(uint8_t rs);
uint8_t lcdReadAddress();
uint8_t lcdReadData();
void lcdInit();
void lcdWriteCommand(uint8_t command);
void lcdWriteData(uint8_t data);
void lcdString(char * buf);

#define LCD_EN 2
#define LCD_RW 1
#define LCD_RS 0

#define LCD_COMMAND_CLEAR 0x01
#define LCD_COMMAND_ENTRY_MODE 0x04
#define LCD_COMMAND_ON 0x08
#define LCD_COMMAND_MOVE 0x10
#define LCD_COMMAND_FUNCTION_SET 0x20
#define LCD_OPTION_8BITS 0x10
#define LCD_OPTION_2LINES 0x08
#define LCD_OPTION_DISPLAY 0x04
#define LCD_OPTION_INCREMENT 0x02

#define LCD_DATA_PORT PORTA
#define LCD_DATA_INPUT PINA
#define LCD_DATA_DDR DDRA
#define LCD_CONTROL_PORT PORTB
#define LCD_CONTROL_DDR DDRB

#define BITS4
```

On the beginning of the file we can see declarations of functions the library provides. Then we have definitions for some commands of LCD protocol and their options. Below them there are definitions of used ports and pins. These *#define* directives allows for some flexibility. However you have to bear in mind that the library assumes that the data pins are 4 more significant pins of a port. If you want to use arbitrary pins you have to modify the library. The last thing in the header file is the definition of *BITS4* macro. With this macro the code is compiled for 4-bit communication. If you remove or comment out this definition then the code will be compiled for 8-bit communication.

There is also the *lcdString()* function which sends characters from provided buffer until it reaches a null character. In other words it displays null-terminated strings, common in C programs.

The header file contains declarations of functions defined in .c file and defines ports and pins used by the library.

Now let's analyze main program:

```
#include "lcd.h"
#include <avr/interrupt.h>
#include <avr/io.h>

char buf[80];
volatile uint8_t cursorState = 0x04;
volatile uint8_t i = 0;

void sendString();

void init_UART(void)
{
        UBRRL = 103; //9600 bps @ 16 MHz crystal
        UCSRB = _BV(RXEN)|_BV(RXCIE); //EX enable, RX int
enable
        UCSRC = _BV(URSEL)|_BV(UCSZ1)|_BV(UCSZ0); //n,8,1
}

ISR(USART_RXC_vect) {
        uint8_t data = UDR;
        switch (data) {
                case 0x01: lcdWriteCommand(LCD_COMMAND_CLEAR);
                        break;
```

```
          case 0x02:
                  cursorState |= 0x02;
                  lcdWriteCommand(LCD_COMMAND_ON |
cursorState);
                  break;
          case 0x03:
                  cursorState &= ~0x02;
                  lcdWriteCommand(LCD_COMMAND_ON |
cursorState);
                  break;
          case 0x04:
                  cursorState |= 0x01;
                  lcdWriteCommand(LCD_COMMAND_ON |
cursorState);
                  break;
          case 0x05:
                  cursorState &= ~0x01;
                  lcdWriteCommand(LCD_COMMAND_ON |
cursorState);
                  break;
          case 0x06:
                  lcdWriteCommand(LCD_COMMAND_MOVE);
                  break;
          case 0x07:
                  lcdWriteCommand(LCD_COMMAND_MOVE | 0x04);
                  break;
          case 0x08:
                  lcdWriteCommand(LCD_COMMAND_MOVE | 0x08);
                  break;
          case 0x09:
                  lcdWriteCommand(LCD_COMMAND_MOVE | 0x08 |
0x04);
                  break;
          case 0xfe:
                  lcdString(buf);
                  i = 0;
                  break;
          default: buf[i++] = data;
      }
      if (data==0xff) {
          i = 0;
          sendString();
      }
```

```
}

int main (void) {
        init_UART();
        sei();
        lcdInit();
        while(1);
        return 0;
}

void sendString() {
        lcdWriteCommand(0x01);
        uint8_t i = 0;
        while (buf[i]!=0) {
                lcdWriteData(buf[i]);
                i++;
        }
        lcdWriteCommand(0xC0);
        i++;
        while (buf[i]!=0xff) {
                lcdWriteData(buf[i]);
                i++;
        }
}
```

The *main()* function initializes USART and enables interrupts. Then it enables interrupts and initializes an LCD. After that program goes into an infinite loop and all things are done in USART interrupt handler.

Looking at the handler we see a protocol used to communicate with outside world. Bytes of values from 1 to 9 are interpreted as commands. There are three LCD commands supported: clearing the display, changing cursor mode and scrolling. As we can see the parameterized LCD commands were translated into unparameterized commands. This makes controlling the display easier. As you can learn from HD44780 documentation the command for display control takes three parameters: enabling/disabling display, enabling/disabling cursor and enabling/disabling blinking. If we want to change only one property we need to know current state and change only one bit. This is why we use the *cursorState* variable. It is initialized to 4 (bit 2 is set) because we want to have the display enabled. When cursor commands come we change only bits 1 and 0 in the byte. Similar situation is in case of moving cursor/scrolling the display but this time we don't need to remember previous value. We just set bits 4 (cursor shift/display scroll) and 3 (left/right).

The most complicated function is displaying text. This is due to the fact that there are two lines on the display. The bytes which are not commands are interpreted as string characters and are stored in a buffer. The end of transmission is signaled by byte of value 255 (FFh). When it is received program sends characters to LCD using *sendString()* function. It moves cursor to first position of upper line and sends characters until it reached null character. Then it moves cursor to the second line and continues sending until byte FFh. This way we can send contents of two lines and have them shown on LCD.

The feature described above is nice but sometimes we would like to send text in parts, what then? It's very simple, just send a null-terminated string followed by FEh byte. After receiving this byte contents of buffer will be sent to LCD without moving cursor.

Presented library, main code and communication protocol are very simple. Some aspects, like displaying custom characters or reading DD RAM and CG RAM, are not presented here. However with information included on the beginning of the chapter they should be easy to implement. You are encouraged to make modifications end experiments. HD44780 is a de facto standard so there is plenty of information available about LCDs with drivers compatible with this chip.

6.8.3. STM32 code

Many parts of the code written for STM32 are identical to the parts of AVR code, the other are similar. Differences are only in low level operations, like reading and writing to ports, configuring ports and performing delays. Let's again start with main LCD library file.

```
#include "lcd.h"

uint8_t switchedTo4BitMode = 0;
uint8_t checkBusyFlag = 0;

void lcdWriteCommand(uint8_t command) {
      if (switchedTo4BitMode){
            checkBusyFlag = 1;
            lcdWrite(command, 0);
            checkBusyFlag = 0;
            lcdWrite(command << 4, 0);
      }else {
            lcdWrite(command, 0);
      }
}

void lcdWriteData(uint8_t data) {
      if (switchedTo4BitMode){
            checkBusyFlag = 1;
            lcdWrite(data, 1);
            checkBusyFlag = 0;
            lcdWrite(data << 4, 1);
      }else {
            lcdWrite(data, 1);
      }
}

uint8_t lcdReadAddress() {
      if (switchedTo4BitMode){
        return ((lcdRead(0) & 0xf0) | (lcdRead(0) >> 4));
      }else{
        return lcdRead(0);
      }
}

uint8_t lcdReadData(){
```

```
      if (switchedTo4BitMode){
        return ((lcdRead(1) & 0xf0) | (lcdRead(1) >> 4));
      }else{
        return lcdRead(1);
      }
}

void lcdWrite(uint8_t command, uint8_t rs) {
      if (checkBusyFlag) while (lcdReadAddress() & 0x80)
Delay(10);

      //R/W low, write mode
      GPIO_ResetBits(LCD_CONTROL_PORT, LCD_RW);
      Delay(10);

      //configure port for write
      GPIO_InitTypeDef GPIO_InitStructure;
#ifdef BITS4
      GPIO_InitStructure.GPIO_Pin = 0xf0;
#else
      GPIO_InitStructure.GPIO_Pin = 0xff;
#endif
      GPIO_InitStructure.GPIO_Speed = GPIO_Speed_2MHz;
      GPIO_InitStructure.GPIO_Mode = GPIO_Mode_Out_PP;
      GPIO_Init(LCD_DATA_PORT, &GPIO_InitStructure);

      //R/S
      if (rs) GPIO_SetBits(LCD_CONTROL_PORT, LCD_RS); else
GPIO_ResetBits(LCD_CONTROL_PORT, LCD_RS);

      //data/command
#ifdef BITS4
      GPIO_ResetBits(LCD_DATA_PORT, ~command & 0xf0);
      GPIO_SetBits(LCD_DATA_PORT, command & 0xf0);
#else
      GPIO_ResetBits(LCD_DATA_PORT, ~command);
      GPIO_SetBits(LCD_DATA_PORT, command);
#endif
      //EN high
      GPIO_SetBits(LCD_CONTROL_PORT, LCD_EN);
      //wait 1 us
      Delay(2);
      //EN low
```

```
        GPIO_ResetBits(LCD_CONTROL_PORT, LCD_EN);
        Delay(2);
}

uint8_t lcdRead(uint8_t rs) {
        //configure port for read
        GPIO_InitTypeDef GPIO_InitStructure;
#ifdef BITS4
        GPIO_InitStructure.GPIO_Pin = 0xf0;
#else
        GPIO_InitStructure.GPIO_Pin = 0xff;
#endif
        GPIO_InitStructure.GPIO_Speed = GPIO_Speed_2MHz;
        GPIO_InitStructure.GPIO_Mode = GPIO_Mode_IN_FLOATING;
        GPIO_Init(LCD_DATA_PORT, &GPIO_InitStructure);

        //R/S
        if (rs) GPIO_SetBits(LCD_CONTROL_PORT, LCD_RS); else
GPIO_ResetBits(LCD_CONTROL_PORT, LCD_RS);
        //R/W high
        GPIO_SetBits(LCD_CONTROL_PORT, LCD_RW);
        //EN high
        GPIO_SetBits(LCD_CONTROL_PORT, LCD_EN);
        //wait 1 us
        Delay(2);
        //command
        uint8_t data = GPIO_ReadInputData(LCD_DATA_PORT);
        //EN low
        GPIO_ResetBits(LCD_CONTROL_PORT, LCD_EN);
        Delay(2);
        return data;
}

void lcdInit() {
        init_GPIO();
        delayMs(15);
        lcdWriteCommand(LCD_COMMAND_FUNCTION_SET |
LCD_OPTION_8BITS); // set interface - 8-bit
        delayMs(4);
        lcdWriteCommand(LCD_COMMAND_FUNCTION_SET |
LCD_OPTION_8BITS); // set interface - 8-bit
        delayMs(1);
```

```
        lcdWriteCommand(LCD_COMMAND_FUNCTION_SET |
LCD_OPTION_8BITS); // set interface - 8-bit
#ifdef BITS4
        delayMs(1);
        lcdWriteCommand(LCD_COMMAND_FUNCTION_SET); // set
interface - 4-bit
        delayMs(1);
        switchedTo4BitMode = 1;
        lcdWriteCommand(LCD_COMMAND_FUNCTION_SET |
LCD_OPTION_2LINES); // set interface - 4-bit, 2 lines, 5x8
font
#else
        checkBusyFlag = 1;
        lcdWriteCommand(LCD_COMMAND_FUNCTION_SET |
LCD_OPTION_8BITS | LCD_OPTION_2LINES); // set interface - 8-
bit, 2 lines, 5x8 font
#endif
        lcdWriteCommand(LCD_COMMAND_ON); // off - nothing to be
on
        lcdWriteCommand(LCD_COMMAND_ON | LCD_OPTION_DISPLAY);
// display on
        lcdWriteCommand(LCD_COMMAND_ENTRY_MODE |
LCD_OPTION_INCREMENT); // scroll
        lcdWriteCommand(LCD_COMMAND_CLEAR); // clear
}

void lcdString(char * buf) {
        while (*buf != 0) {
                lcdWriteData(*buf);
                buf++;
        }
}

void init_GPIO() {
  RCC_APB2PeriphClockCmd(RCC_APB2Periph_GPIOA |
RCC_APB2Periph_GPIOB, ENABLE);

  GPIO_InitTypeDef GPIO_InitStructure;
  GPIO_InitStructure.GPIO_Speed = GPIO_Speed_2MHz;
  GPIO_InitStructure.GPIO_Mode = GPIO_Mode_Out_PP;
  GPIO_InitStructure.GPIO_Pin = LCD_EN | LCD_RW | LCD_RS;
  GPIO_Init(LCD_CONTROL_PORT, &GPIO_InitStructure);
}
```

```
void Delay(volatile unsigned int delay)
{
  for(;delay>0;delay--);
}

void delayMs(uint8_t d) {
  for (uint8_t i = 0; i < d; i++) Delay(12000);
}
```

As it was said differences aren't big. Besides of different I/O functions for ports there is not much to talk about. The delay is performed using *Delay()* function presented before. There is additional *delayMs()* function which is a wrapper for *Delay()* providing millisecond delays. The value 12 000 comes from the fact that the microcontroller works at 72 MHz and iteration of a loop takes 6 clock cycles. Therefore we need 12 000 iteration per millisecond.

Another difference is *init_GPIO()* function. On AVR initialization of control pins is as outputs is one line. On ARM, with the peripheral library, it's a little bit more. For example we have to initialize clocks of used I/O ports. Hence this part of code has been moved to a separate function.

The library's header file is very similar to the one written for AVR. An interesting thing is that in the following code we use pin 2 of port B while previous chapter discouraged using this pin as it servers as BOOT1 too. Here however we are driving an LCD which has a little bit different internal circuit comparing to transistor based driver for LED displays and there is no issue with pin 2 of port B.

```
#include <stm32f10x_gpio.h>

void lcdWrite(uint8_t command, uint8_t rs);
void lcdWriteCommand(uint8_t command);
void lcdWriteData(uint8_t data);
uint8_t lcdRead(uint8_t rs);
uint8_t lcdReadAddress();
uint8_t lcdReadData();
void lcdInit();
void lcdString(char * buf);
void delayMs(uint8_t i);
void Delay(unsigned int delay);
void init_GPIO();

#define LCD_EN GPIO_Pin_2
#define LCD_RW GPIO_Pin_1
#define LCD_RS GPIO_Pin_0

#define LCD_COMMAND_CLEAR 0x01
#define LCD_COMMAND_ENTRY_MODE 0x04
#define LCD_COMMAND_ON 0x08
#define LCD_COMMAND_MOVE 0x10
#define LCD_COMMAND_FUNCTION_SET 0x20
#define LCD_OPTION_8BITS 0x10
#define LCD_OPTION_2LINES 0x08
#define LCD_OPTION_DISPLAY 0x04
#define LCD_OPTION_INCREMENT 0x02

#define LCD_DATA_PORT GPIOA
#define LCD_CONTROL_PORT GPIOB

#define BITS4
```

The code of main program is also very similar to the ARM version. The main difference is of course in USART initialization.

```
#include "stm32f10x.h"
#include "lcd.h"

void init_USART();
void sendString();
```

```c
char buf[80];
volatile uint8_t cursorState = 0x04;
volatile uint8_t i = 0;

int main(void)
{
  init_USART();
  lcdInit();
  while(1);

}

void sendString() {
      lcdWriteCommand(0x01);
      uint8_t i = 0;
      while (buf[i]!=0) {
              lcdWriteData(buf[i]);
              i++;
      }
      lcdWriteCommand(0xC0);
      i++;
      while (buf[i]!=0xff) {
              lcdWriteData(buf[i]);
              i++;
      }
}

void USART1_IRQHandler(void)
{
  if (USART_GetITStatus(USART1, USART_IT_RXNE) != RESET)
  {
    uint8_t data = USART_ReceiveData(USART1);
    switch (data) {
          case 0x01: lcdWriteCommand(LCD_COMMAND_CLEAR);
                    break;
          case 0x02:
                    cursorState |= 0x02;
                    lcdWriteCommand(LCD_COMMAND_ON |
cursorState);
                    break;
          case 0x03:
                    cursorState &= ~0x02;
```

```
                        lcdWriteCommand(LCD_COMMAND_ON |
cursorState);
                        break;
            case 0x04:
                        cursorState |= 0x01;
                        lcdWriteCommand(LCD_COMMAND_ON |
cursorState);
                        break;
            case 0x05:
                        cursorState &= ~0x01;
                        lcdWriteCommand(LCD_COMMAND_ON |
cursorState);
                        break;
            case 0x06:
                        lcdWriteCommand(LCD_COMMAND_MOVE);
                        break;
            case 0x07:
                        lcdWriteCommand(LCD_COMMAND_MOVE | 0x04);
                        break;
            case 0x08:
                        lcdWriteCommand(LCD_COMMAND_MOVE | 0x08);
                        break;
            case 0x09:
                        lcdWriteCommand(LCD_COMMAND_MOVE | 0x08 |
0x04);
                        break;
            case 0xfe:
                        lcdString(buf);
                        i = 0;
                        break;
            default: buf[i++] = data;
        }
        if (data==0xff) {
            i = 0;
            sendString();
        }
    }
}

void init_USART(){

  RCC_APB2PeriphClockCmd(RCC_APB2Periph_GPIOA |
RCC_APB2Periph_USART1, ENABLE);
```

```
GPIO_InitTypeDef GPIO_InitStructure;
GPIO_InitStructure.GPIO_Pin = GPIO_Pin_9;
GPIO_InitStructure.GPIO_Speed = GPIO_Speed_2MHz;
GPIO_InitStructure.GPIO_Mode = GPIO_Mode_AF_PP;
GPIO_Init(GPIOA, &GPIO_InitStructure);

GPIO_InitStructure.GPIO_Pin = GPIO_Pin_10;
GPIO_InitStructure.GPIO_Speed = GPIO_Speed_2MHz;
GPIO_InitStructure.GPIO_Mode = GPIO_Mode_IN_FLOATING;
GPIO_Init(GPIOA, &GPIO_InitStructure);

USART_InitTypeDef USART_InitStructure;
USART_InitStructure.USART_BaudRate = 9600;
USART_InitStructure.USART_WordLength = USART_WordLength_8b;
USART_InitStructure.USART_StopBits = USART_StopBits_1;
USART_InitStructure.USART_Parity = USART_Parity_No;
USART_InitStructure.USART_HardwareFlowControl =
USART_HardwareFlowControl_None;
USART_InitStructure.USART_Mode = USART_Mode_Tx |
USART_Mode_Rx;
USART_Init(USART1, &USART_InitStructure);
USART_ITConfig(USART1, USART_IT_RXNE, ENABLE);
USART_Cmd(USART1, ENABLE);

NVIC_PriorityGroupConfig(NVIC_PriorityGroup_1);
NVIC_InitTypeDef NVIC_InitStructure;
NVIC_InitStructure.NVIC_IRQChannel = USART1_IRQn;
NVIC_InitStructure.NVIC_IRQChannelSubPriority = 1;
NVIC_InitStructure.NVIC_IRQChannelPreemptionPriority = 1;
NVIC_InitStructure.NVIC_IRQChannelCmd = ENABLE;
NVIC_Init(&NVIC_InitStructure);

}
```

6.8.4. PC code

An example C# code is presented below. It is extremely simple so there should be no problems with understanding it.

```
using System;
using System.IO.Ports;
```

```
using System.Windows.Forms;

namespace SerialSample8
{
    public partial class Form1 : Form
    {
        SerialPort ComPort = new SerialPort("COM1");

        public Form1()
        {
            InitializeComponent();
            if (ComPort.IsOpen) ComPort.Close();
            ComPort.BaudRate = 9600;
            ComPort.Parity = Parity.None;
            ComPort.StopBits = StopBits.One;
            ComPort.DataBits = 8;
            ComPort.Handshake = Handshake.None;
            ComPort.Open();
        }

        private void btnClear_Click(object sender, EventArgs
e)
        {
            ComPort.Write(new byte[] { 0x01 }, 0, 1);
        }

        private void btnSendText_Click(object sender,
EventArgs e)
        {
            ComPort.Write(txtLine1.Text);
            ComPort.Write(new byte[] { 0x00 }, 0, 1);
            ComPort.Write(txtLine2.Text);
            ComPort.Write(new byte[] { 0xff }, 0, 1);
        }

        private void btnCursorOn_Click(object sender,
EventArgs e)
        {
            ComPort.Write(new byte[] { 0x02 }, 0, 1);
        }

        private void btnCursorOff_Click(object sender,
EventArgs e)
```

```
        {
            ComPort.Write(new byte[] { 0x03 }, 0, 1);
        }

        private void btnBlinkOn_Click(object sender,
EventArgs e)
        {
            ComPort.Write(new byte[] { 0x04 }, 0, 1);
        }

        private void btnBlinkOff_Click(object sender,
EventArgs e)
        {
            ComPort.Write(new byte[] { 0x05 }, 0, 1);
        }

        private void btnCursorLeft_Click(object sender,
EventArgs e)
        {
            ComPort.Write(new byte[] { 0x06 }, 0, 1);
        }

        private void button1_Click(object sender, EventArgs
e)
        {
            ComPort.Write(new byte[] { 0x07 }, 0, 1);
        }

        private void btnScrollLeft_Click(object sender,
EventArgs e)
        {
            ComPort.Write(new byte[] { 0x08 }, 0, 1);
        }

        private void btnScrollRight_Click(object sender,
EventArgs e)
        {
            ComPort.Write(new byte[] { 0x09 }, 0, 1);
        }

        private void btnAppend_Click(object sender, EventArgs
e)
        {
```

```
        ComPort.Write(txtLine1.Text);
        ComPort.Write(new byte[] { 0x00 }, 0, 1);
        ComPort.Write(new byte[] { 0xfe }, 0, 1);
      }
    }
}
```

Besides of serial port initialization, almost all the code are just handlers of click events of several buttons placed on the main window. Most of them just send one-byte commands. Two of them deal with sending text from text boxes according to the protocol described earlier.

The one thing which needs our attention is that the *Write()* method which takes string type by default sends only standard ASCII characters. This means that characters of codes greater than 127 are changed to question marks. Thus if you want to send extended ASCII characters use array of bytes. You can also change the encoding but this would require modification of microcontroller's code.

6.9. LED & PWM

I/O ports of microcontrollers allow generating digital signals. This is perfect in many situations but sometimes we need an analog signal. Some microcontrollers have built-in digital-to-analog converters and the situation is pretty easy. We can also use a DAC in the form of separate chip or even make it from scratch from handful of resistors.

But there is also a quite different approach: use PWM, which stands for pulse-width modulation. The idea is to use an I/O pin to generate a square wave of rather high frequency and control duration of high and low states. By changing how long there is zero and one on the pin we can control average voltage on it. The higher the duration of one with respect to zero the higher the voltage. Let's say VCC is 5 volts while period of one is 10 milliseconds and period of zero is 25 ms. Hence the average voltage is 40% of VCC which gives 2 V.

But still we are talking about square wave. Although it's not a problem in many situations, sometimes we would like to have constant voltage. A simple solution is to use a low-pass filter consisting of a resistor and a capacitor. The voltage on the resistor will equal to average voltage on the I/O pin generating the square wave. Of course this solution is not perfect because the output voltage is not ideally constant, there is still a small AC signal of the original frequency. Nonetheless this solution is suitable in most cases. For better filtering you can make a few things:

- use capacitor of bigger value

- increase frequency

- draw as small current as possible from the filter

- use resistor of bigger value

You must remember however that the bigger the resistance the lower the current which you can take from the filter because resistance in the filter and of the load create a voltage divider. Additionally the bigger the RC time constant the slower is the filter's response. This can be important if we want to make fast changes of the voltage level provided through PWM.

In following example we want to control LED's brightness so no RC filter is required, there is just a LED and standard current limiting resistor.

6.9.1. Hardware configuration

As it was said above all we need is a LED diode and a suitable resistor. You can choose if you want to connect the diode between microcontroller's pin and GND or between pin and VCC. In the first case brightness will be proportional to period of one and in the second to period of zero. The following code assumes the first case. Generally it's the same circuit as in examples 1 and 2. For AVR the difference is that we cannot use arbitrary pin but have to use OC0 (Timer/Counter0 Output Compare Match Output) pin. In ATmega32 this pin shares its function with PB3 so we have to connect the LED with pin 4 (PB3/OC0/AIN1). In STM32 example we use channel 1 of timer 2 which is shares its function with pin 0 of port A. Therefore the circuit is exactly the same as in examples 1 and 2.

Serial port and microcontrollers

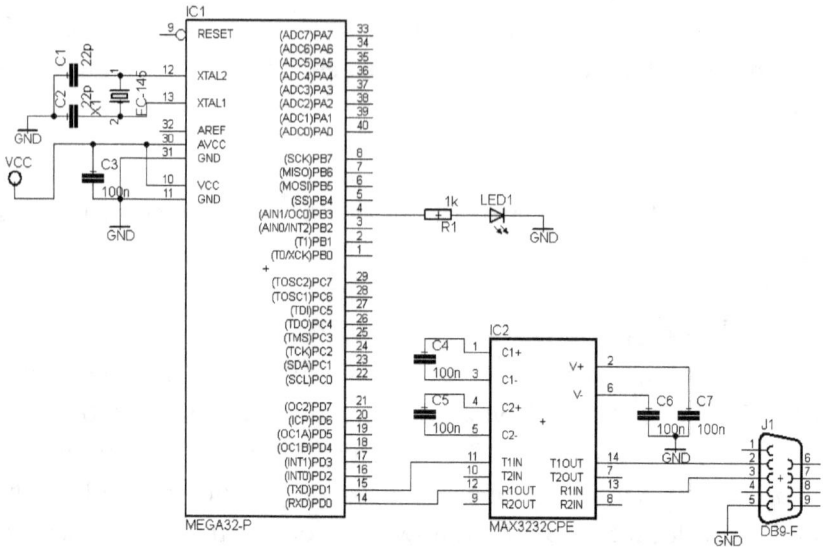

Fig. 6.9.1.1. Circuit with ATmega32.

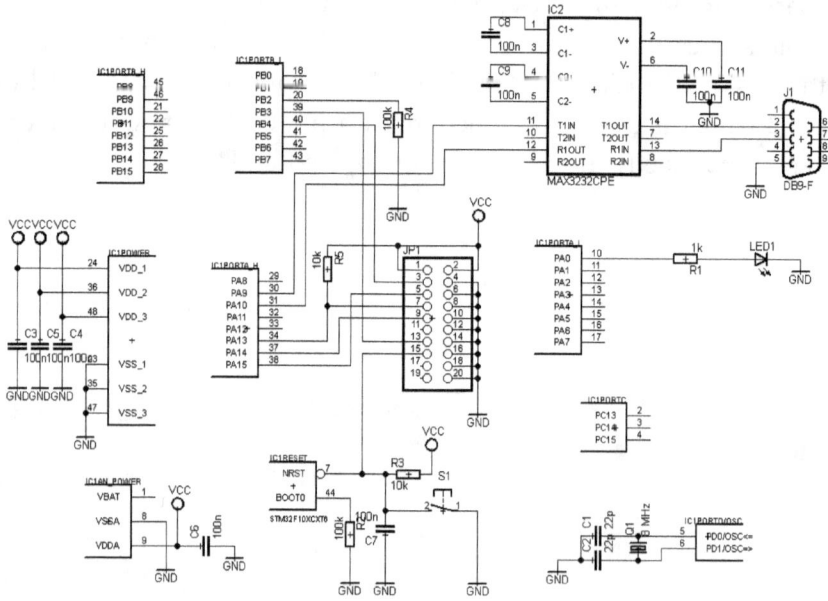

Fig. 6.9.1.2. Circuit with STM32F103CBT6.

6.9.2. AVR code

This time the code is pretty simple:

```c
#include <avr/io.h>
#include <avr/interrupt.h>

void init_PWM() {
        TCCR0 = _BV(WGM01)|_BV(WGM00)|_BV(COM01)|_BV(CS00);   //
Fast PWM Mode, CLK/1
        DDRB = _BV(DDB3);
}

void init_UART(void)
{
        UBRRL = 103; //9600 bps @ 16 MHz crystal
        UCSRB = _BV(RXEN)|_BV(RXCIE); //EX enable, RX int
enable
        UCSRC = _BV(URSEL)|_BV(UCSZ1)|_BV(UCSZ0); //n,8,1
}

ISR(USART_RXC_vect) {
        OCR0 = UDR;
}

int main() {
        init_PWM();
        init_UART();
        sei();
        OCR0 = 0;
        while(1);
        return 0;
}
```

The most important thing is actually one line, the one where we set value of *TCCR0* register which controls Timer0. By setting *WGM01* and *WGM00* bits we enable so called Fast PWM Mode. In this mode the timer counts from 0 to 255 and then starts again from 0. Each time it is incremented it is compared against *OCD0* register. What happens on match is determined by the *COM01* and *COM00*. Here only *COM01* is set which means, that when the counter is set to 0 the output pin is set to one. Later, when match occurs, pin is set to zero. This way how long the pin is in high state depends on *OCD0*.

Initially *OCD0* is set to 0 so we start with the LED completely turned off. Of course you can select different initial value. The register is updated when there is a byte received from USART.

6.9.3. STM32 code

The code for ARM is longer but it works in the same way, using one of available timers.

```c
#include "stm32f10x.h"

void init_USART();
void init_Timer();
void sendString();

int main(void)
{
   init_USART();
   init_Timer();
   while(1);

}

void USART1_IRQHandler(void)
{
        if (USART_GetITStatus(USART1, USART_IT_RXNE) != RESET)
        {
                uint8_t data = USART_ReceiveData(USART1);
                TIM_SetCompare1(TIM2, data);
        }
}


void init_USART(){

  RCC_APB2PeriphClockCmd(RCC_APB2Periph_GPIOA |
RCC_APB2Periph_USART1, ENABLE);
```

```
  GPIO_InitTypeDef GPIO_InitStructure;
  GPIO_InitStructure.GPIO_Pin = GPIO_Pin_9;
  GPIO_InitStructure.GPIO_Speed = GPIO_Speed_2MHz;
  GPIO_InitStructure.GPIO_Mode = GPIO_Mode_AF_PP;
  GPIO_Init(GPIOA, &GPIO_InitStructure);

  GPIO_InitStructure.GPIO_Pin = GPIO_Pin_10;
  GPIO_InitStructure.GPIO_Speed = GPIO_Speed_2MHz;
  GPIO_InitStructure.GPIO_Mode = GPIO_Mode_IN_FLOATING;
  GPIO_Init(GPIOA, &GPIO_InitStructure);

  USART_InitTypeDef USART_InitStructure;
  USART_InitStructure.USART_BaudRate = 9600;
  USART_InitStructure.USART_WordLength = USART_WordLength_8b;
  USART_InitStructure.USART_StopBits = USART_StopBits_1;
  USART_InitStructure.USART_Parity = USART_Parity_No;
  USART_InitStructure.USART_HardwareFlowControl =
USART_HardwareFlowControl_None;
  USART_InitStructure.USART_Mode = USART_Mode_Tx |
USART_Mode_Rx;
  USART_Init(USART1, &USART_InitStructure);
  USART_ITConfig(USART1, USART_IT_RXNE, ENABLE);
  USART_Cmd(USART1, ENABLE);

  NVIC_PriorityGroupConfig(NVIC_PriorityGroup_1);
  NVIC_InitTypeDef NVIC_InitStructure;
  NVIC_InitStructure.NVIC_IRQChannel = USART1_IRQn;
  NVIC_InitStructure.NVIC_IRQChannelSubPriority = 1;
  NVIC_InitStructure.NVIC_IRQChannelPreemptionPriority = 1;
  NVIC_InitStructure.NVIC_IRQChannelCmd = ENABLE;
  NVIC_Init(&NVIC_InitStructure);

}

void init_Timer(){

  RCC_APB1PeriphClockCmd(RCC_APB1Periph_TIM2, ENABLE);

  TIM_TimeBaseInitTypeDef TIM_InitStruct;
  TIM_InitStruct.TIM_ClockDivision = TIM_CKD_DIV1;
  TIM_InitStruct.TIM_CounterMode = TIM_CounterMode_Up;
  TIM_InitStruct.TIM_Period = 255;
  TIM_InitStruct.TIM_Prescaler = 100;
```

```
  TIM_TimeBaseInit(TIM2, &TIM_InitStruct);

  TIM_OCInitTypeDef TIM_OCInitStructure;
  TIM_OCInitStructure.TIM_OCMode = TIM_OCMode_PWM1;
  TIM_OCInitStructure.TIM_OutputState =
TIM_OutputState_Enable;
  TIM_OCInitStructure.TIM_Pulse = 0;
  TIM_OCInitStructure.TIM_OCPolarity = TIM_OCPolarity_High;
  TIM_OC1Init(TIM2, &TIM_OCInitStructure);

  TIM_Cmd(TIM2, ENABLE);

  GPIO_InitTypeDef GPIO_InitStructure;
  GPIO_InitStructure.GPIO_Pin = GPIO_Pin_0;
  GPIO_InitStructure.GPIO_Speed = GPIO_Speed_2MHz;
  GPIO_InitStructure.GPIO_Mode = GPIO_Mode_AF_PP;
  GPIO_Init(GPIOA, &GPIO_InitStructure);
}
```

USART is configured in a standard way. The structure for timer initialization is also filled in a familiar way. There is no clock division, the timer counts up. Period is set to 255 because the PC application sends single byte each time and of course byte's maximal value is 255. Prescaler is set to 100 but it can be 10 or 1000 as well, in case of LED it doesn't matter much. But if you set prescaler to a high value, like 50 000, blinks are slow enough that you can see them.

When the basic timer parameters are configured we set up one of its channels to work in PWM mode. There are two PWM modes: 1 and 2, they differ in polarity. In mode 1, with *TIM_OCPolarity* set to *TIM_OCPolarity_High*, the timer starts with high state on its output. When timer's counter reaches value defined by *TIM_Pulse* then the output is switched to low state. This way the greater the *TIM_Pulse* value the longer the time of high state. In other words average voltage on timer's output grows with growing value of *TIM_Pulse*. In PWM mode 2 or with *TIM_OCPolarity* set to *TIM_OCPolarity_Low* the situation is opposite, average voltage decreases while *TIM_Pulse* increases. So in fact you can affect the way the timer generates PWM signal in three ways: by changing PWM mode, by changing polarity, or by making it count up or down.

TIM_Pulse is set to 0 so we start with LED turned off. Because we want to use generated PWM signal we set *TIM_OutputState* to *TIM_OutputState_Enable*. When the output compare structure is ready we use it to configure channel 1 of timer 2. The last thing is to configure appropriate GPIO pin to work as timer's output. The pin which works with channel 1 of timer 2 is pin 1 of port A. Thus we set its mode as alternate function with push-pull.

The rest is trivial. When USART1 interrupt happens, we read received value and set it as a compare value for channel 1 of timer 2. It is possible to do it by filling the structure for *TIM_OC1Init()* function, writing the received value to the *TIM_Pulse* field and calling *TIM_OC1Init()*, just as we did during initialization. But it would be on overkill. Just take a look at definition of this function. It's quite long and everything we need to do is to set value of timer's *CCR1* register. Thus we use *TIM_SetCompare1()* function. We don't waste processor time for structure initialization and configuring register we have already configured.

6.9.4. PC code

The code is extremely simple, we are just sending value of the slider when it is dragged.

```
using System;
using System.IO.Ports;
using System.Windows.Forms;

namespace SerialSample9
{
    public partial class Form1 : Form
    {
        SerialPort ComPort = new SerialPort("COM1");

        public Form1()
```

```
        {
            InitializeComponent();
            if (ComPort.IsOpen) ComPort.Close();
            ComPort.BaudRate = 9600;
            ComPort.Parity = Parity.None;
            ComPort.StopBits = StopBits.One;
            ComPort.DataBits = 8;
            ComPort.Handshake = Handshake.None;
            ComPort.Open();
        }

        private void trackBar1_Scroll(object sender,
EventArgs e)
        {
            ComPort.Write(new byte[] { (byte)trackBar1.Value
}, 0, 1);
        }
    }
}
```

6.10. Receiving RC-5 commands

RC-5 is a very popular standard for infra-red remote control. It was developed by Philips and is now used by many other manufacturers of consumer electronics. The protocol is quite simple and thus easy to implement. It's just a little bit more complicated than RS-232. First of all, RC-5 uses carrier frequency which is 36 kHz. This means, that each bit consists of series of pulses of infra-red light. That's first difference to RS-232. In that standard nothing really happened during transmission of a bit. The voltage was remaining on the same level from beginning of a bit to its end. In RC-5 the situation is different, a transmitter is constantly blinking with infra-red light and there are 32 pulses for each bit. If we have pulses for each bit so how do we know ones from zeros? That's the second difference: RC-5 makes use of so called Manchester coding. In this coding we don't transmit bits as states but as transitions. As I said, there are 32 pulses in each bit. But they are present only in half of a bit. Which half? That's the issue. If nothing happens in first half and pulses are present in the second one then logic one is transmitted. If pulses are in half of a bit but not in the second then logic zero is transmitted. So we talk about logic one when in a bit there is transition from lack of pulses to their presence, about zero in opposite situation. Or, if we are dealing only with a receiver, we can just look at the second part of each transmitted bit. If there are pulses we have one, if there are none than we have zero.

In this chapter we are going to talk about receiving RC-5 commands, sent using regular RC-5 remote. If you are going to buy such a remote, make sure it supports this protocol. As for the receiver, there are many integrated infra-red receivers, suitable for RC-5. One of the most popular such devices is TFMS5360. ZL15AVR development board uses different element: TSOP31236, but they are practically identical. They have the same package, pinout and operate in the same way. These integrated receivers make building devices very easy. You don't need to build special analog circuit for infra-red photo-diode and you don't have to care about the carrier frequency. You just get logic zero when 36 kHz infra-red pulses are transmitted and logic one when idle. Thus your code for microcontroller has to take care about Manchester coding and analyze consecutive bits, it doesn't have to detect 36 kHz carrier frequency.

Let's back to the protocol, but at higher level. There are 14 bits in each transmission. The first bit is a start bit and marks beginning of a transmission and is always 1. The second bit is called field bit and it indicates if transmitted command belongs to lower field (commands 0 – 63) or upper field (commands 64 – 127). Lower field is marked by 1, higher by 0. Originally second bit worked as additional start bit and was always 1. Philips changed it when it became clear that 63 commands are not enough. Field bit can be therefore described as 7th command bit, but negated. Bit 3 is a control bit. It says if a button on remote control was pressed just now or is still being pressed from the time of last transmission and now is being repeated. It allows a receiver to distinguish between two or more subsequent button presses or one long press. Every consecutive press causes the control bit to be toggled. If user is pressing a button continuously then the bit remains unchanged. Bits from 4 to 8 are address bits. Many remotes can control more than one device and address allows distinguishing them. The last 6 bits (9 – 14) are command bits and tell us what button is pressed. Both address and command are encoded from the most significant bit to the least significant.

6.10.1.   Hardware configuration

TFMS5360 is a popular IR receiver of signals having 36 kHz carrier frequency. There are many other integrated receivers compatible with this one, e.g. TSOP31236. It has just three pins: GND, VS (+5V power supply), and OUT. As this element requires 5V supply you have to provide it in your circuit. If your microcontroller is powered from 5V supply (like ATmega32) then there is no problem. For STM32 you need to provide both voltages as those chips are usually powered from 3.3 V supply. In such cases the circuit as a whole is powered from 5V supply but this voltage goes just to elements which require it, for example the IR receiver we are talking about. It also goes to a 3.3 voltage regulator which powers the microcontroller. A popular 3.3 V voltage regulator is LM1117-3.3. Depending on a manufacturer prefix can be different, e.g. LD or REG. Suffix denotes voltages so it has to be 3.3.

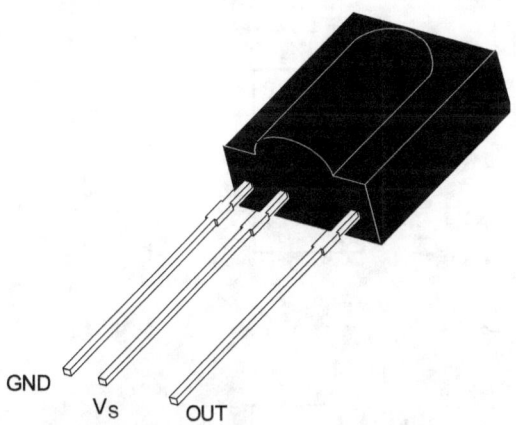

GND

Vs

OUT

Fig. 6.10.1.1. TFMS5360.

The circuit is simple. We just need to connect output of IR receiver to a pin which our code is going to read. For both AVR and STM32 we use pin 0 of port A. It's also good to add simple RC circuit for filtering supply voltage for the IR receiver.

Serial port and microcontrollers

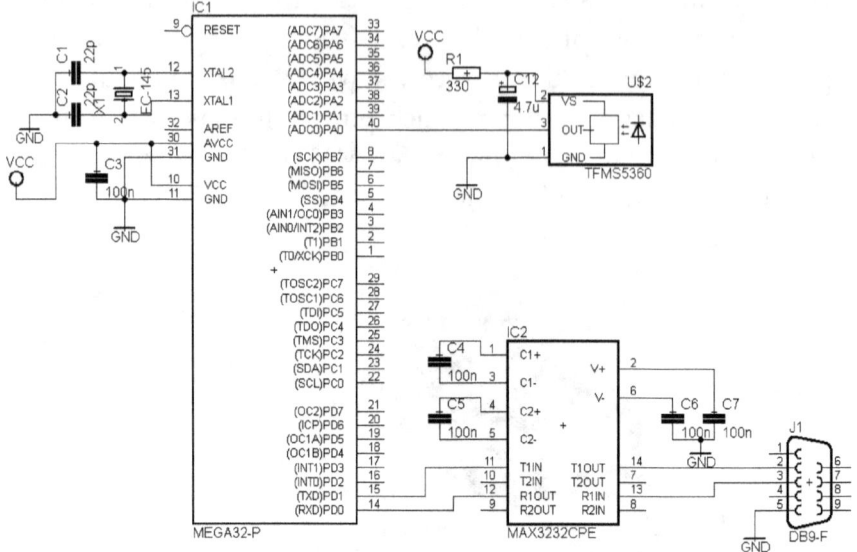

Fig. 6.10.1.2. Circuit with ATmega32.

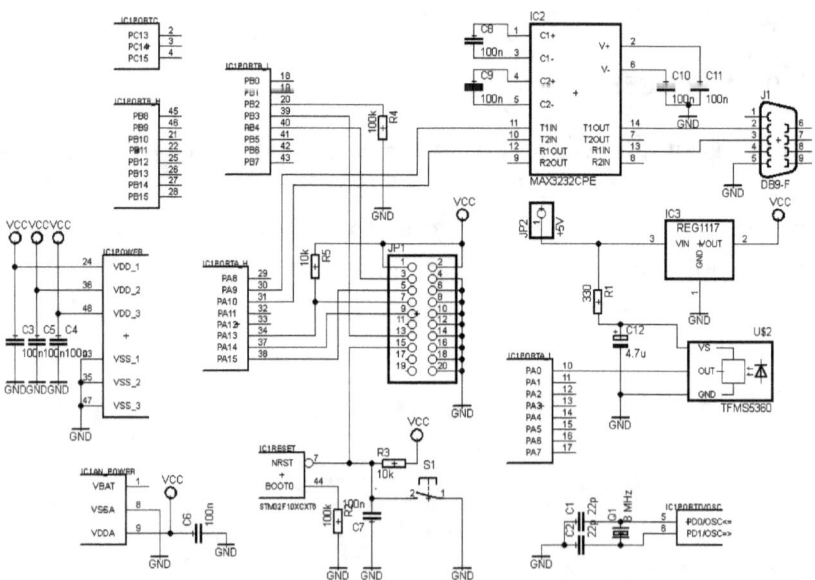

Fig. 6.10.1.3. Circuit with STM32F103CBT6.

6.10.2.   AVR code

The source code may seem a little bit complicated at first but it's really quite simple.

```c
#include <avr/io.h>
#include <avr/interrupt.h>
#define F_CPU 16000000UL
#include <util/delay.h>

uint8_t waitAndRead(uint8_t withTimeout);
void init_UART();
void init_Timer();
void sendByte(uint8_t c);
volatile uint8_t timeout = 0;

void init_UART(void)
{
        UBRRL = 103; //9600 bps @ 16 MHz crystal
        UCSRB = _BV(TXEN); //EX enable
        UCSRC = _BV(URSEL)|_BV(UCSZ1)|_BV(UCSZ0); //n,8,1
}

void init_Timer() {
        TIMSK = _BV(TOIE0);             // Int T0 Overflow enabled
        TCCR0 = _BV(CS00)|_BV(CS01);    // CLK/64
}

ISR(TIMER0_OVF_vect)
{
        timeout = 1;
}

int main() {
        init_UART();
        init_Timer();
        sei();
        uint8_t oldControl = 0;
        while(1) {
                uint8_t ret = 0;
                uint16_t val = 0;
                for (uint8_t i=0; i<14; i++) {
                        val <<= 1;
```

```
                    ret = waitAndRead(i != 0);
                    if (ret == 2) break;
                    val |= ret;
            }
            if (ret == 2) continue;
            uint8_t command = val & 0b00111111;
            uint8_t address = (val >> 6) & 0b00011111;
            uint8_t control = (val >> 11) & 1;
            sendByte(command);
            sendByte(address);
            sendByte(control != oldControl);
            oldControl = control;
        }
        return 0;
}

uint8_t waitAndRead(uint8_t withTimeout) {
        _delay_ms(1);
        uint8_t state = PINA & _BV(PINA0);
        uint8_t oldState = state;
        if (withTimeout) {
                TCNT0 = 0;
                timeout = 0;
                while ((state == oldState) && (timeout == 0))
state = PINA & _BV(PINA0);
                if (timeout == 1) return 2;
        } else {
                while (state == oldState) state = PINA &
_BV(PINA0);
        }
        if (state) return 0; else return 1;
}

void sendByte(uint8_t c) {
        while(!(UCSRA & _BV(UDRE)));
        UDR = c;
}
```

As it was said in the theoretical introduction, for easy decoding of Manchester coding it is enough to look at the second half of each bit. How to detect second half? By waiting for a change of state which occurs between the halves. OK, but state doesn't occur only at this situation. Solution is very simple and is implemented using *waitAndRead()* function. Let's assume that we are in idle state, nothing is transmitted. The function starts and it waits for 1 millisecond. This delay doesn't anything now but will be important later. After the delay we check state of the pin connected to IR receiver and start waiting for it to change in a *while* loop. First bit comes and we read its value. Then the function is called again by the *for* loop located in *main* function. Now we see what the 1 ms delay is for. Thanks to this delay we skip any changes of state which can happen between second half of first bit and first half of second bit. When the delay is over we are right in the second half of current bit. In this way we can read consecutive bits and return them to *for* loop in *main()*. This loop stores received bits in a variable. When all 14 bits are read then we extract command, address and control bit. It's done simply with bit shifts and logical AND. When we have all those three data we can send them through serial port. The case of control bit is a little bit different. We don't send it as it is but compare it with its previous variable. If there was a change, which means that user has just pressed the button and is not pressing it continuously, we transmit 1. Otherwise we transmit 0. 0 and 1 values comes from != operator which compares variables.

Basically that's all. But for sure you noticed that a timer is used in our code and you probably wonder what we need it for. Well, you can remove call to *init_Timer()* and see for yourself but chances are that you won't see any difference. OK, I'm rushing to explain. Have you wondered what happens if there is some error during transmission? Maybe the batteries are almost dead in your remote or someone was passing by between transmitter and receiver. What will happen then? The code would wait for remaining bits of transmitted message until a new message is transmitted. Then it will treat some bits from the new message as belonging to the old one. In the end you will observe a bit shift in received data and you will have to reset your microcontroller. That's very bad from user experience point of view.

Hence simple solution which makes use of a timer. When we wait for second half of a bit we not only wait for change of state but also check for timeout. If we see that we wait too long we return a special value. In our example it's 2. When this value is returned then we stop waiting for next bits and start over and wait for the first bit. As a timer we use Timer0 which is 8-bit. In our code it is configured with prescaler set to 64 so for 16 MHz clock timer's counter is incremented 250 000 times per second. This means that it overflows in a little bit less than a millisecond, about 976 microseconds. This is just a little bit more than the time of half of a bit which is 889 microseconds. Before we start waiting for a bit we set timer's counter to zero and the *timer* variable as well. Then we wait for change of state or for *timer* variable to be changed by the timer's overflow interrupt. One more thing. We need to check for timeout for all bits but the first one. Therefore *waitAndRead()* takes a parameter which tells whether we are dealing with the first bit or not.

6.10.3.   STM32 code

The code for ARM STM32 microcontroller is as follows.

```
#include "stm32f10x.h"

uint8_t waitAndRead(uint8_t withTimeout);
void init_UART();
void init_Timer();
void sendByte(uint8_t c);
volatile uint8_t timeout = 0;
void delayMs(uint8_t d);

void init_UART(){

  RCC_APB2PeriphClockCmd(RCC_APB2Periph_GPIOA |
RCC_APB2Periph_USART1, ENABLE);

  GPIO_InitTypeDef GPIO_InitStructure;
  GPIO_InitStructure.GPIO_Pin = GPIO_Pin_9;
  GPIO_InitStructure.GPIO_Speed = GPIO_Speed_2MHz;
  GPIO_InitStructure.GPIO_Mode = GPIO_Mode_AF_PP;
```

```
  GPIO_Init(GPIOA, &GPIO_InitStructure);

  GPIO_InitStructure.GPIO_Pin = GPIO_Pin_0;
  GPIO_InitStructure.GPIO_Mode = GPIO_Mode_IN_FLOATING;
  GPIO_Init(GPIOA, &GPIO_InitStructure);

  USART_InitTypeDef USART_InitStructure;
  USART_InitStructure.USART_BaudRate = 9600;
  USART_InitStructure.USART_WordLength = USART_WordLength_8b;
  USART_InitStructure.USART_StopBits = USART_StopBits_1;
  USART_InitStructure.USART_Parity = USART_Parity_No;
  USART_InitStructure.USART_HardwareFlowControl =
USART_HardwareFlowControl_None;
  USART_InitStructure.USART_Mode = USART_Mode_Tx |
USART_Mode_Rx;
  USART_Init(USART1, &USART_InitStructure);
  USART_Cmd(USART1, ENABLE);

}


void init_Timer(){

      RCC_APB1PeriphClockCmd(RCC_APB1Periph_TIM2, ENABLE);

      TIM_TimeBaseInitTypeDef TIM_InitStruct;
      TIM_InitStruct.TIM_ClockDivision = TIM_CKD_DIV1;
      TIM_InitStruct.TIM_CounterMode = TIM_CounterMode_Up;
      TIM_InitStruct.TIM_Period = 9;
      TIM_InitStruct.TIM_Prescaler = 7199;
      TIM_TimeBaseInit(TIM2, &TIM_InitStruct);

      TIM_ClearFlag(TIM2, TIM_FLAG_Update);
      TIM_ITConfig(TIM2, TIM_IT_Update, ENABLE);
      TIM_Cmd(TIM2, ENABLE);

      NVIC_InitTypeDef NVIC_InitStruct;
      NVIC_InitStruct.NVIC_IRQChannel = TIM2_IRQn;
      NVIC_InitStruct.NVIC_IRQChannelCmd = ENABLE;
      NVIC_InitStruct.NVIC_IRQChannelPreemptionPriority = 0;
      NVIC_InitStruct.NVIC_IRQChannelSubPriority = 0;
      NVIC_Init(&NVIC_InitStruct);
}
```

```
void TIM2_IRQHandler() {
      if (TIM_GetITStatus(TIM2, TIM_IT_Update)!=RESET) {
            TIM_ClearFlag(TIM2, TIM_FLAG_Update);
            timeout = 1;
      }
}


int main() {
      init_UART();
      init_Timer();
      uint8_t oldControl = 0;
      while(1) {
            uint8_t ret = 0;
            uint16_t val = 0;
            for (uint8_t i=0; i<14; i++) {
                  val <<= 1;
                  ret = waitAndRead(i != 0);
                  if (ret == 2) break;
                  val |= ret;
            }
            if (ret == 2) continue;
            uint8_t command = val & 0x3f;
            uint8_t address = (val >> 6) & 0x1f;
            uint8_t control = (val >> 11) & 1;
            sendByte(command);
            sendByte(address);
            sendByte(control != oldControl);
            oldControl = control;
      }

}

uint8_t waitAndRead(uint8_t withTimeout) {
      delayMs(1);
      uint8_t state = GPIO_ReadInputDataBit(GPIOA,
GPIO_Pin_0);
      uint8_t oldState = state;
      if (withTimeout) {
            TIM_SetCounter(TIM2, 0);
            timeout = 0;
```

```
            while ((state == oldState) && (timeout == 0))
state = GPIO_ReadInputDataBit(GPIOA, GPIO_Pin_0);
            if (timeout == 1) return 2;
        } else {
            while (state == oldState) state =
GPIO_ReadInputDataBit(GPIOA, GPIO_Pin_0);
        }
        if (state) return 0; else return 1;
}

void sendByte(uint8_t c) {
        while(USART_GetFlagStatus(USART1, USART_FLAG_TXE) ==
RESET);
        USART_SendData(USART1, c);
}

void delayMs(uint8_t d) {
        TIM_SetCounter(TIM2, 0);
        timeout = 0;
        while (timeout == 0);
}
```

We use the same algorithm as for AVR. Let's look at differences in implementation. In our code we use interrupts for a timer and for USART. Therefore we need interrupt handlers and code which sets interrupt vectors to point at those handlers. For this purpose we use startup file *startup_stm32f10x_md.s* provided with the SMT32 Standard Peripheral Library. The feature of this file is that it not only sets interrupt vectors but also calls *SystemInit()* which is defined in *system_stm32f10x.c* file. This file changes clock source to high speed external (HSE) and sets PLL multiplier factor to 9 so the microcontroller works with 72 MHz clock. Of course you don't need to work at 72 MHz. You can have your own startup file or you can change the *SystemInit()* function to set different frequency. But here we stick with original library files and adapt our code to 72 MHz frequency. *SystemInit()* is called just after reset, before *main()*.

The TIM2 timer is configured in a standard way. Its prescaler is set to 7199 which means that main frequency is divided by 7 200. Period is set to 9 so the timer's counter counts from 0 to 9 which gives 10 states. Hence the frequency is 72 MHz / 72 000 = 1 kHz. This way we get 1 millisecond delay. The timer is used for both busy waiting type delay (the *delayMs()* function) and for timeout implementation.

The rest of a program, including serial communication, is realized in a standard manner.

6.10.4.   PC code

```
using System;
using System.Windows.Forms;
using System.IO.Ports;

namespace SerialSample10
{
    public partial class Form1 : Form
    {
        SerialPort ComPort = new SerialPort("COM1");
        int command, address, control;
        public Form1()
        {
            InitializeComponent();
            ComPort.BaudRate = 9600;
            ComPort.Parity = Parity.None;
            ComPort.StopBits = StopBits.One;
            ComPort.DataBits = 8;
            ComPort.Handshake = Handshake.None;
            ComPort.DataReceived += OnSerialDataReceived;
            ComPort.Open();
        }

        private void OnSerialDataReceived(object sender,
SerialDataReceivedEventArgs args)
        {
            command = ComPort.ReadByte();
            address = ComPort.ReadByte();
            control = ComPort.ReadByte();
            lblCommand.Invoke(new
EventHandler(lblCommand_DataReceived));
            lblAddress.Invoke(new
EventHandler(lblAddress_DataReceived));
```

```
        lblControl.Invoke(new
EventHandler(lblControl_DataReceived));
        if (command == 12) MessageBox.Show("Power
pressed!");
    }

    private void lblCommand_DataReceived(object sender,
EventArgs e)
    {
        lblCommand.Text = command.ToString();
    }

    private void lblAddress_DataReceived(object sender,
EventArgs e)
    {
        lblAddress.Text = address.ToString();
    }

    private void lblControl_DataReceived(object sender,
EventArgs e)
    {
        if (control != 0) lblControl.Text = "New press";
else lblControl.Text = "Continuous press";
    }
  }
}
```

The code presented above is very simple and is very similar to previous examples. Received values are just displayed on appropriate labels. Additionally the code checks is command 12 was received which is for power button. In such case a message box is displayed. Of course you can add more actions, for example run some application, move the window, play some sound, change volume, etc.

6.11.   Using Bluetooth modules

Bluetooth is a popular standard of wireless communication. You can find it in devices like mobile phones, hands free headsets, keyboards, mouses, computers, any many other devices. You can use Bluetooth in your on projects too. While it's a quite complicated standard you don't need to know much about it though. There are integrated Bluetooth modules which make adding wireless functionality to your devices a very easy task.

One of popular such modules is BTM-222. It's a Class 1 module which means that it can operate on ranges up to about 100 meters. It supports Serial Port Profile (SPP) which can be used to emulate a serial cable. In module's datasheet you can find information about other interfaces, like USB and PCM, however they are not supported by BTM-222 firmware.

From the point of view of a microcontroller the module works as a standard serial device using *TX* and *RX* lines. At the PC side the Bluetooth driver creates a virtual COM port which can be used by applications, similarly as for RS-232-to-USB adapters. Bluetooth modules work usually at 3.3V, you have to keep it in mind if your device is powered from supply of other voltage. You may need to add a voltage regulator (like LD1117DT33) and some buffer (for example 74V1G126). A Bluetooth module can work with an antenna to achieve longer transmission distance. You can connect a specialized antenna, like TCA10FR or just a piece of a wire. Bluetooth frequency is 2.4 GHz so for half-wave antenna the wire should be 6 centimeters long and for quarter-wave it should be 3 centimeters long.

When the BTM-222 module is powered on it is in command mode. In this mode it waits for configuration commands on its serial line. They are called AT commands and each one of them starts with letters "AT". The "AT" itself comes from the word "attention". Using AT commands you can change name of the module, scan for nearby devices, connect to one of them, change PIN or switch to master mode. For example *ATL1* changes baud rate to 9600 bps and ATL? Inquires currently set speed. To list all settings send *ATI1* command. AT commands have to be finished with carriage return (CR) character, which is byte of value 13 (0Dh). When you write code for a microcontroller to send AT commands to Bluetooth modules you need to introduce short delay after each character. It should be about 5 ms if echo is disabled and 25 ms if enabled. Of course in data mode no delay is needed, you can send data with full speed of module's serial interface (by default 19200 bps). Some modules, including BTM-222, can go into sleep mode. It takes place in master mode, when auto connecting is disabled, the module is in command mode and there was no AT command for 60 seconds. When in sleep mode, the module would ignore first character of received AT command. Therefore if there is a possibility that the module can enter sleep mode, you have to wake it up before issuing AT commands. For example you can send a CR character.

By default the module works in slave mode. Master device is the device which initiates connection and controls parameters of Bluetooth transmission. If your device is meant to work with a computer it will probably be better to have the module in slave mode. In case the device needs to scan for other devices and choose the one to connect you need master mode. In slave mode a microcontroller which is connected to the Bluetooth module doesn't have to do much. It just waits for a connection. When the connection is established the module sends "CONNECT" string, followed by address of the other device. The whole string is preceded and followed by carriage return (CR) and line feed (LF) characters. From this moment the module works as a transparent link between the attached microcontroller and the other side, e.g. PC application.

A (establish a connection)	Works in master mode, establishes a connection. In slave mode the command will be rejected.	
	Modifiers	Description
	A	Connect to a Bluetooth device (It's only available when an address was assigned with „ATD=xxxxxxxxxxx")
	A1 – A8	Connect to a Bluetooth neighborhood device 1~8 (ATF? Result)
B (Display local BD address)	This command displays the local device BD address	
	Modifiers	Description
	B?	Inquire the Local BD address
D (Set Remote BD address)	For security purpose we can specify the unique remote device that can be connected. In master role, it automatically inquires and searches the slave even if the slave is undiscoverable. In slave role, the command should be used as a filter condition to accept master's inquiry.	
	Modifiers	Description
	D=xxxxxxxxxxx	"xxxx-xx-xxxxxx" is 12 digit hex symbol
	D0 (Default)	Clear Remote BD address setting, inquire any slave in master mode or accept any master in slave mode.
	D?	Inquire the Remote BD address setting
E	This command specifies whether the device	

(Local Echo)	should echo characters received from the UART back to the DTE/DCE.	
	Modifiers	Description
	E0	Command characters received from the UART are not echoed back to the DTE/DCE.
	E1 (Default)	Command characters received from the UART are echoed back to the DTE/DCE.
	E?	Inquire the current setting
F (Find Bluetooth device)	This command is used to find any Bluetooth device in neighborhood within 60 seconds timeout. If any device is found, its name and address will be listed. The search ends with a message „Inquiry ends, xx device(s) found." This command is available only when the adaptor is in the master role.	
	Modifiers	Description
	F?	Inquire scan Bluetooth neighborhood devices.
H (Discoverable Control)	This command specifies whether the device could be discovered by remote master device. note : waitting for 15 seconds afert ATH1 command to take the effect	
	Modifiers	Description
	H0	The device enters undiscoverable mode. If a pair has been established, the original connection can be connected again. But other remote master device cannot discover this

		device.
	H1 (Default)	The device enters discoverable mode.
	H?	Inquire the current setting
I (Information)	This command is used to Inquiry the F/W version	
	Modifiers	Description
	I?	Inquire the version Codes
K (Stop bits setting)	This command is used to specify one or two stop bits of COM port	
	Modifiers	Description
	K0 (Default)	One Stop bit
	K1	Two stop bits
	K?	Inquire the current setting
L (Baud Rate Control)	This command is used to specify the baud rate of COM port	
	Modifiers	Description
	L0	4800bps
	L1	9600bps
	L2 (Default)	19200bps
	L3	38400bps
	L4	57600bps
	L5	115200bps
	L6	230.4Kbps
	L7	460.8Kbps
	L?	Inquire the current setting
M (Parity bits	This command is used to specify the parity bit setting of COM port	

setting)	Modifiers	Description
	M0 (Default)	No Parity bit.
	M1	Odd parity setting.
	M2	Even parity setting
	M?	Inquire the current setting
N (Set device name)	We can specify the device's friendly name using 0 to 9, A to Z, a to z, space and –, which are all valid characters. Note that firs space or -, last space or – isn't permitted. The default name is „Serial Adaptor"	
	Modifiers	Description
	N=xxxxx	"xxxxx" is a character string, maximal length is 16
	N?	Inquire the device name
O (Auto connect setting)	When it's in master mode the command is used to enable/disable auto-connection feature. When it's in slave mode, the command will be rejected.	
	Modifiers	Description
	O0 (Default)	Automatically connects to a device which is assigned in "ATD" or any available device if „ATD" was not assigned.
	O1	Disable auto-connection feature, user should manually use „ATA" command to connect a remote device.

	O?	Inquire the current setting
P (Set PIN code)	This command specifies the PIN number. It allows to disable the PIN code authorization to allow establishing a connection without PIN code. Default PIN number is "1234"	
	Modifiers	Description
	P=xxxx	"xxxx" is 4~8 digit string
	P0	Turn off the PIN code authorization
	P?	Inquire the current PIN number
Q (Result Code Suppression)	The command is used to determine if result codes should be sent to the DTE/DCE. When result codes are suppressed, the device does not generate any characters in response to the completion of a command or when an event occurs. Four result codes : OK, CONNECT, DISCONNECT, ERROR	
	Modifiers	Description
	Q0 (Default)	The device will send Result Codes to the DTE/DCE.
	Q1	The device will not send Result Codes to the DTE/DCE.
	Q?	Inquire the current setting
R (Set Role)	This command specifies whether the device is a master or slave device. If changing the role, the module will warm start and clear all paired addresses.	

	Modifiers	Description
	R0	The device works as master.
	R1 (Default)	The device works as slave.
	R?	Inquire the current setting
U (F/W upgrade)	This command will prompt „Enter DFU mode, Are you sure (y/n)?" message, then press Y to confirm the command. Then you should connect USB cable to PC and run DFU wizard (for DFU wizard please contact us www.rayson.com)	
	Modifiers	Description
	U=password	Password = RaysonUpgrade, go to Upgrade F/W Mode
Z (Application setting)	Restore application settings and warm start.	
	Modifiers	Description
	Z0	Restore factory default setting (19200bps, slave, etc.)
	Z?	Inquire the current setting

During connection the module can be switched to command mode by sending "+++" escape string. These three characters can't be sent immediately one after another, there must be a one second delay between them. When the module switches to command mode it sends "OK" with CR. Now you can issue AT commands while the connection still remains active. To close the connection use ATH command, to return to transparent data mode use ATO command. Disconnection is signaled in similar way as connection, there is

"DISCONNECT" word followed by address of the device which has been disconnected.

You need to take into account two things about the escape string. First, it is sent to the other device. Therefore that device needs to recognize it and ignore. Second, you have to be sure that in any case you won't be sending data which look like the escape string. If it happened then following data will be interpreted as AT commands and won't be transmitted. Therefore you have to think for a while if you need to switch to command mode at all. If not then you can disable the escape string with ATX0 command and then you don't have to worry about what the data you send look like. If you need switching to command mode then you need to analyze what you send and how you send. As it was said earlier, the characters in the escape sequence need to be separated by a 1 second delay. This means that even if it happens that you send "+++" data but without long delays then there is no problem.

BTM-222 and many other Bluetooth modules keep all the settings in non-volatile memory. Therefore you need to perform configuration only once. There is no need to reconfigure a module every time it is powered on. You can implement code for configuration in your microcontroller and run it for example when a button is pressed. You can also use your computer and appropriate adapter, e.g. MAX3232.

It's now a good time to say a few words about connecting a module to a computer. In general we want to connect our module to a microcontroller but for test and configuration purposes we can connect it to a computer. That's the first thing to do before you start writing code for a microcontroller. The BTM-222, like many other modules, has a form of a PCB without any pins. It just has soldering points on sides. Hence for experiments you need a special PCB to easily access the module. You can make such a PCB by yourself or buy a Bluetooth development board like KAmodBTM222.

Fig. 6.11.1. KAmodBTM222 module

This board can be powered from 3.3V or 5V supply thanks to a voltage regulator, you just need to set a jumper accordingly. It has also a buffer which makes it suitable for working with 5V microcontrollers. KAmodBTM222 can be easily connected to microcontroller development boards, such as ZL15AVR and ZL30ARM. ZL15AVR operates at 5V so you need set the jumper to 5V on the KAmodBTM222 and connect its VCC pin to +5V pin on ZL15AVR. In case of ZL30ARM you can set the jumper to 3V or 5V but you need to provide appropriate supply voltage using +3,3V or +5V pins on the ZL30ARM accordingly. Then you can connect RXD and TXD lines. If you want to connect the Bluetooth module to a microcontroller, connect these lines to appropriate microcontroller's pins. But for experiments you may wish to use development board's RS-232 voltage converter to connect the module to a computer. In this case connect TXD with TxD and RXD with RxD. And don't forget about the ground (GND). Connect the development board with your computer using RS-232 cable, turn on the development board and open your favorite terminal program, e.g. PuTTY. Set the speed to 19200 bps, no parity, one stop bit, 8 bits of data, and no flow control.

Serial port and microcontrollers

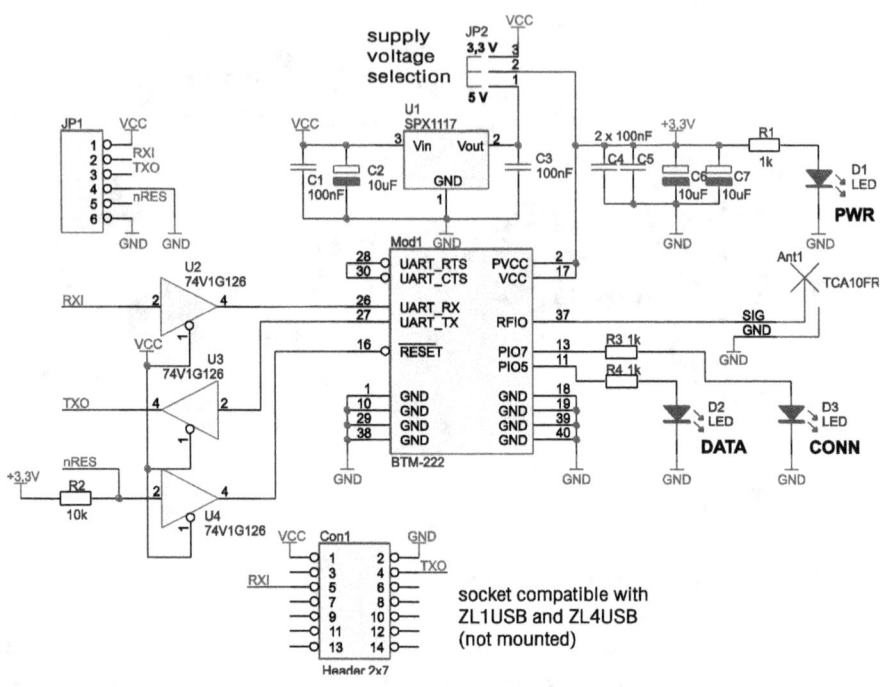

Fig. 6.11.2. KAmodBTM222 circuit

Now everything is ready, you can type AT commands. For start you can type "ATI0" or "ATI1" to check version number or current settings. By default echo is enabled so you can see what you type. It's nice during tests but if the module is going to be connected to a microcontroller then your code has to be prepared for the fact that all command will be echoed back. Or you can disable echo. On your computer launch scan for Bluetooth devices and perform the pairing procedure. The default PIN for BTM-222 is 1234. When the computer and the module are paired together check which virtual COM ports have been created for the module. Here they are COM15 and COM16. One of them is for incoming connections (initiated by the module when it works in master mode and the computer is slave) and the other is for outgoing connections (initiated by computer when it works in master mode and the module is slave).

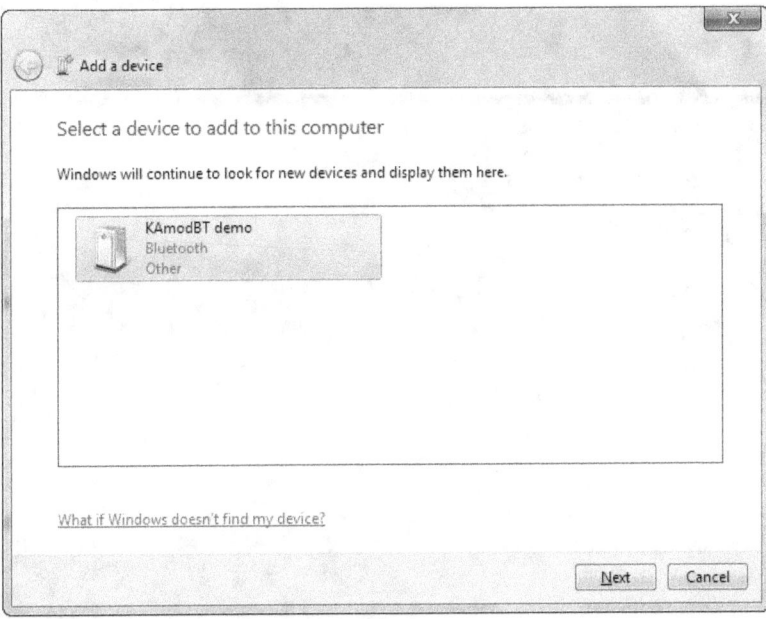

Fig. 6.11.3. Searching for Bluetooth devices

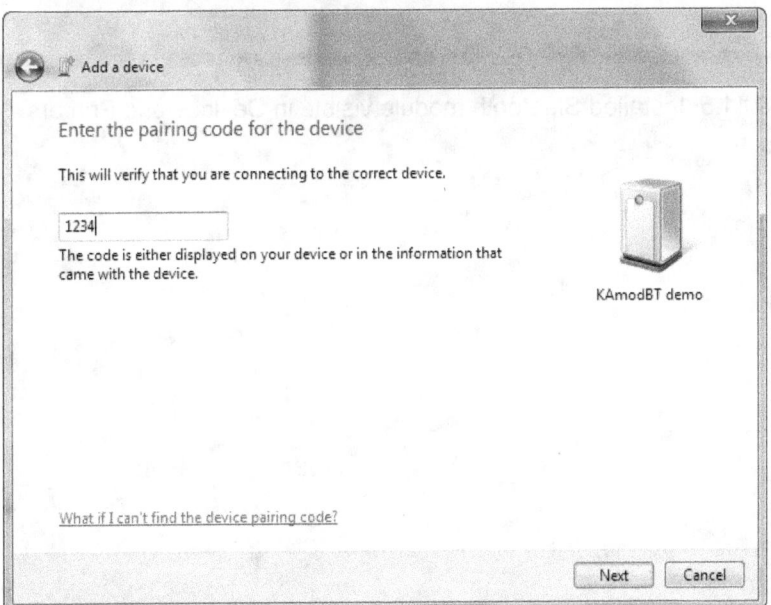

Fig. 6.11.4. Entering pairing code.

Fig. 6.11.5. Installed Bluetooth module visible in Devices and Printers

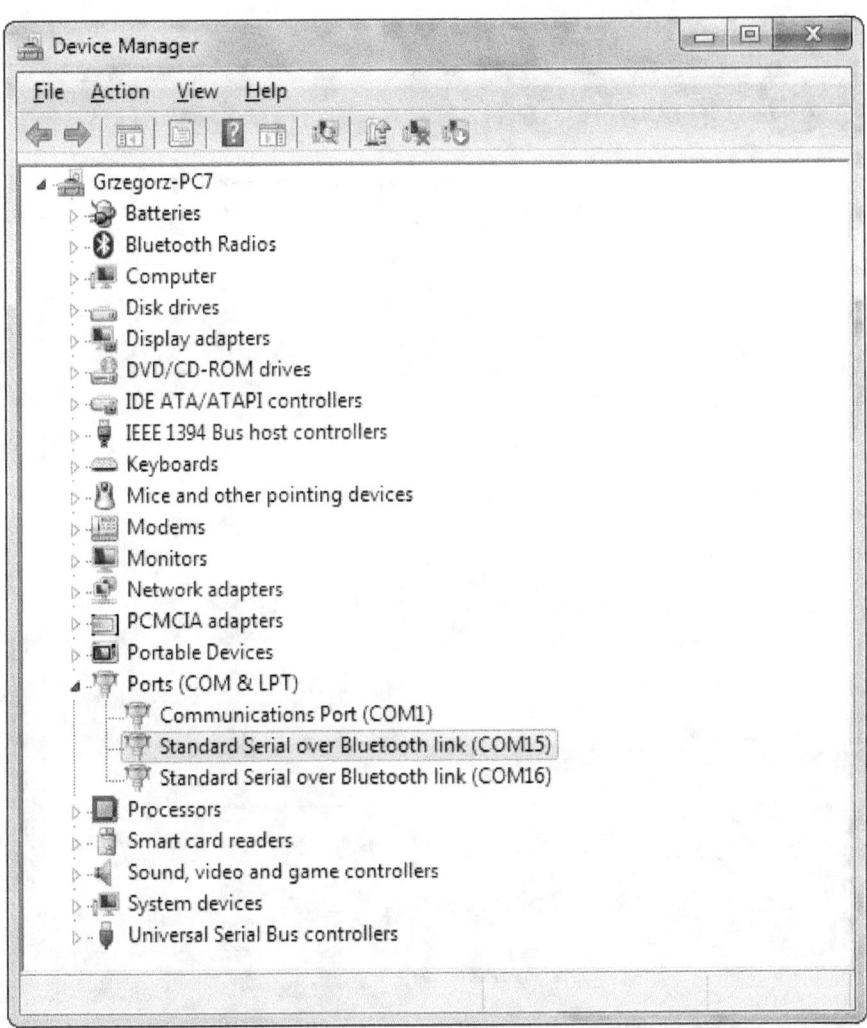

Fig. 6.11.6. Bluetooth serial ports visible in Device Manager.

Serial port and microcontrollers

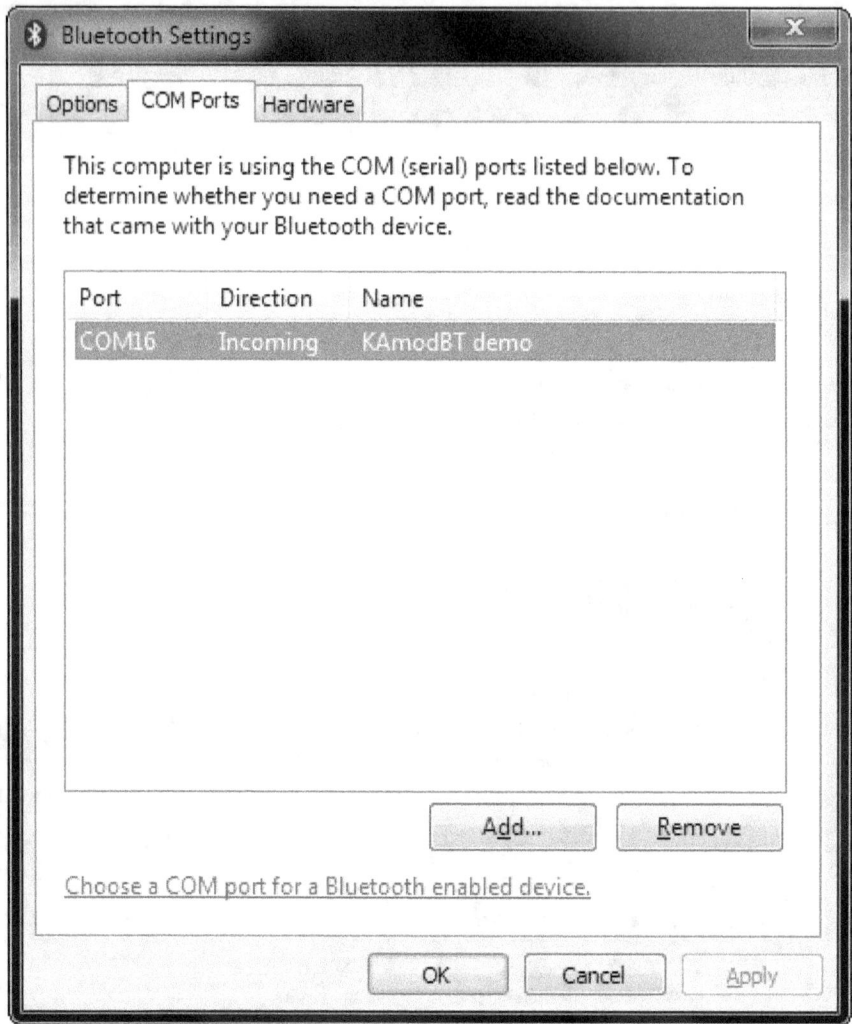

Fig. 6.11.7. Incoming serial port shown in Bluetooth Settings

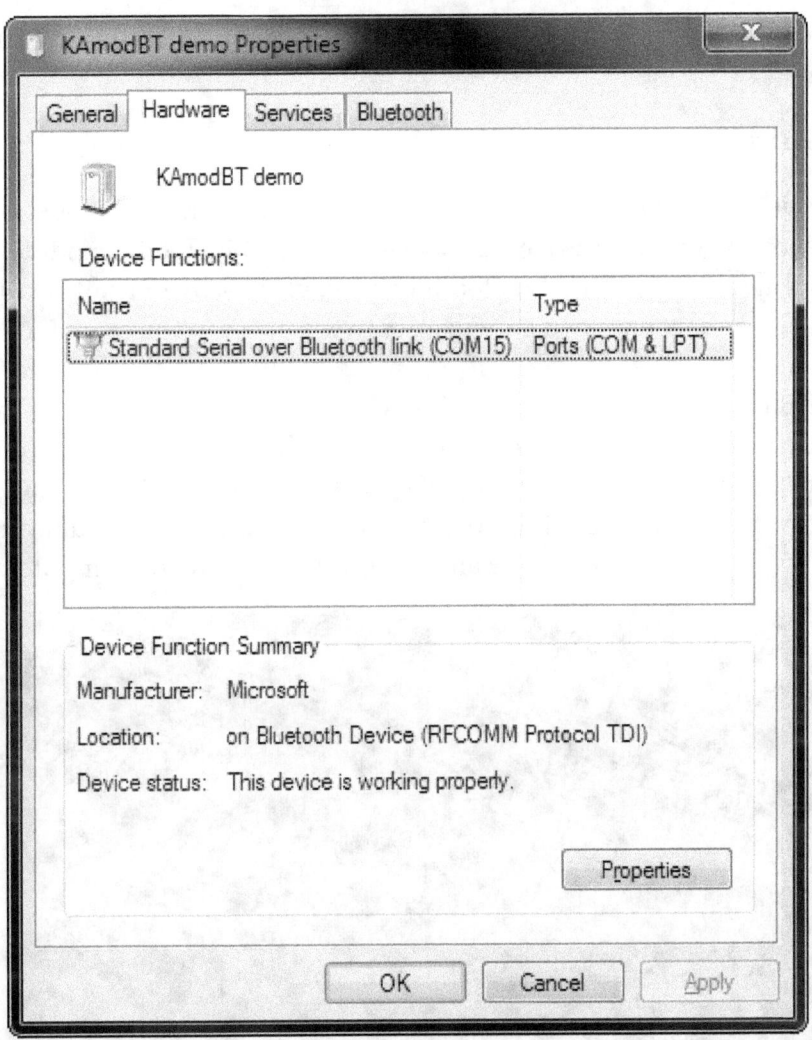

Fig. 6.11.8. Outgoing port shown in module's properties.

Now you can run terminal application and open this port. On the terminal working with the module you should see now the "CONNECT" string followed by address of your computer. Now you can type on both terminals and the text should go to the other one by Bluetooth. If everything is working at this point you can connect Bluetooth module to a microcontroller and start writing code for it.

Remark

By default "CONNECT" and "DISCONNECT" strings are enabled. Hence a microcontroller connected to a Bluetooth module is informed whether the Bluetooth connection is active or not. On the other hand it has to distinguish between these statuses and other data. You have to keep it in mind while writing code for a microcontroller. You can also disable statuses using ATQ0 command. Then the microcontroller won't know whether the connection is established or not but actually it's not always required. You have to make a decision according to a protocol you use for communication between the microcontroller and application running on computer.

Fig. 6.11.9. Terminal of a Bluetooth module. Configuration check and incoming connection.

Fig. 6.11.10. Terminal running on a computer.

Screenshots 6.11.9 and 6.11.10 show an example session. The first one shows communication with Bluetooth module using its serial interface. Here the screenshot is from PuTTY running on a PC connected with a module using low voltage USB-serial interface (CDC-232). First we issue ATI0, ATN?, and ATP? Commands to check firmware version, name, and PIN. Then Bluetooth outgoing serial port is opened as shown on second screenshot. This initiates Bluetooth connection which is signaled by the module by „CONNECT" message and MAC address of the Bluetooth interface of the computer which initiated the connection. Then two words are typed: „Hello ", and „word!". First one is typed on a PC and is displayed by the module's terminal. The second one is typed on module's terminal and is visible on a PC.

6.11.1.   Hardware configuration

In this example we use a LED and a button as in our basic examples. The difference is that we are going to use a Bluetooth module instead of a voltage converter (MAX3232). Notice that you have to make a cross-connection between TXD and RXD pins of your microcontroller and of the module. The module and the microcontroller work here as two DTEs.

Most probably you will be experimenting with an integrated Bluetooth development board, like KAmodBTM222, not bare module (BTM-222 or similar). Therefore such configuration is shown below. Anyway you have to take care about proper supply voltage for the module. If you are going to use KAmodBTM222 then set the JP2 jumper to a position appropriate for supply voltage of your microcontroller. In case of ATmega32 it's usually 5V. STM32 is usually powered from 3.3V supply. If you use ZL30ARM development board then you have there both voltages available on the pins of the board. Set the jumper with respect to the voltage you connected to VCC pin of Bluetooth development board. If you are not going to use integrated Bluetooth development board but just the module itself add proper voltage regulator and buffers if your microcontroller is powered from 5V supply.

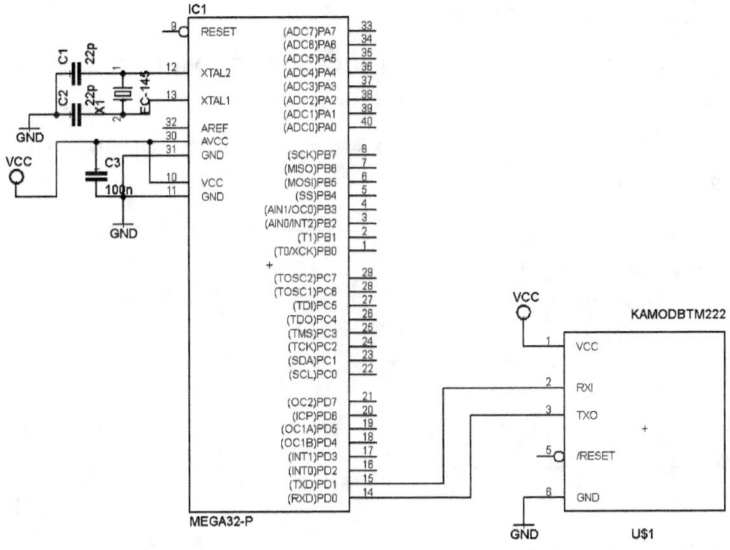

Fig. 6.11.1.1. Circuit with ATmega32.

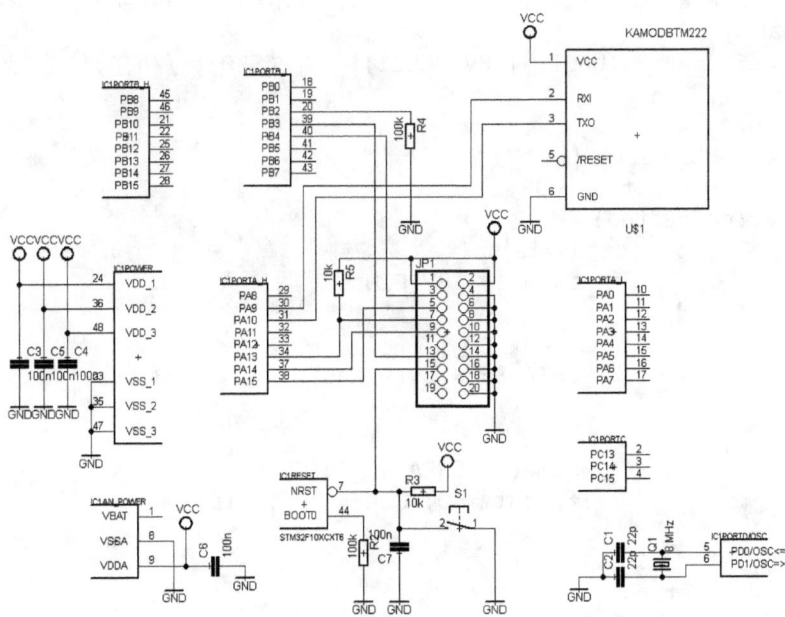

Fig. 6.11.1.2. Circuit with STM32F103CBT6.

6.11.2. AVR code

The code presented here is a modification of simple example program which showed how to handle two-way communication. The novelty is addition of detecting establishing and closing of a Bluetooth connection.

```
#include <avr/io.h>
#include <avr/interrupt.h>
#define F_CPU 16000000U
#include <util/delay.h>

volatile uint8_t state;
volatile uint8_t connected = 0;
volatile uint8_t firstLf = 0;
volatile uint8_t firstChar = 0;

void init_UART (void)
{
      UBRRL = 51; //19200 bps @ 16 MHz crystal
      UCSRB = _BV(RXEN)|_BV(TXEN)|_BV(RXCIE); //EX enable, RX
int enable
      UCSRC = _BV(URSEL)|_BV(UCSZ1)|_BV(UCSZ0); //n,8,1
}

ISR(USART_RXC_vect) {
      uint8_t data = UDR;
      if(connected) {
            if (data=='a') {
                  PORTA |= _BV(PINA0);
            }
            if (data=='b') {
                  PORTA &= ~_BV(PINA0);
            }
            if (data=='r') {
                  while(!(UCSRA & _BV(UDRE)));
                  if(state==0) UDR = 'P'; else UDR = 'N';
            }
      }
      if(firstLf) {
            if(!firstChar) firstChar = data;
      }
      if (data=='\n'){
```

```
            if (firstLf) {
                    firstLf = 0;
                    if (firstChar == 'C') {
                            connected = 1;
                    }
                    if (firstChar == 'D') {
                            connected = 0;
                    }
                    firstChar = 0;
            } else firstLf = 1;
        }
}

int main() {
        uint8_t stateOld = 3;
        init_UART();
        sei();
        DDRA |= _BV(DDA0)|_BV(DDA2);
        PORTA |= _BV(PINA1);
        while(1){
                while(!connected);
                PORTA |= _BV(PINA2);
                while(connected) {
                        state = PINA & _BV(PINA1);
                        if (state!=stateOld) {
                                stateOld = state;
                                while(!(UCSRA & _BV(UDRE)));
                                if(state==0) UDR = 'P'; else UDR =
'N';
                                _delay_ms(50);
                        }
                }
                PORTA &= ~_BV(PINA2);
        }
}
```

The code is quite simple. It makes use of the fact that "CONNECT" and "DISCONNECT" status strings are preceded and followed by <CR><LF>. First the program waits for the leading line feed character. When it is received the *firstLf* variable is set to 1. When next character is received it is saved if that variable is set. This way we store first letter of the status string. It is "C" for "CONNECT" and "D" for "DISCONNECT". When the second <LF> character comes we check the first letter of the status and set the *connected* variable accordingly. The code ignores <LF> characters and doesn't verify other status letters. If you wish you can improve detection of connection state but in most cases it shouldn't be necessary.

6.11.3.   STM32 code

The program for STM32 executes the same algorithm as the program for AVR.

```
#include "stm32f10x.h"

volatile uint8_t state = 0;
volatile uint8_t connected = 0;
volatile uint8_t firstLf = 0;
volatile uint8_t firstChar = 0;


void Delay(unsigned int delay)
{
  for(;delay>0;delay--);
}


void init_UART() {

      RCC_APB2PeriphClockCmd(RCC_APB2Periph_GPIOA |
RCC_APB2Periph_USART1, ENABLE);

      GPIO_InitTypeDef GPIO_InitStructure;
      GPIO_InitStructure.GPIO_Pin = GPIO_Pin_9;
      GPIO_InitStructure.GPIO_Speed = GPIO_Speed_2MHz;
      GPIO_InitStructure.GPIO_Mode = GPIO_Mode_AF_PP;
      GPIO_Init(GPIOA, &GPIO_InitStructure);

      GPIO_InitStructure.GPIO_Pin = GPIO_Pin_10;
```

```
        GPIO_InitStructure.GPIO_Mode = GPIO_Mode_IN_FLOATING;
        GPIO_Init(GPIOA, &GPIO_InitStructure);

        GPIO_InitStructure.GPIO_Pin = GPIO_Pin_1;
        GPIO_InitStructure.GPIO_Mode = GPIO_Mode_IPU;
        GPIO_Init(GPIOA, &GPIO_InitStructure);

        GPIO_InitStructure.GPIO_Pin = GPIO_Pin_0 | GPIO_Pin_2;
        GPIO_InitStructure.GPIO_Mode = GPIO_Mode_Out_PP;
        GPIO_Init(GPIOA, &GPIO_InitStructure);

        USART_InitTypeDef USART_InitStructure;
        USART_InitStructure.USART_BaudRate = 19200;
        USART_InitStructure.USART_WordLength =
USART_WordLength_8b;
        USART_InitStructure.USART_StopBits = USART_StopBits_1;
        USART_InitStructure.USART_Parity = USART_Parity_No;
        USART_InitStructure.USART_HardwareFlowControl =
USART_HardwareFlowControl_None;
        USART_InitStructure.USART_Mode = USART_Mode_Tx |
USART_Mode_Rx;
        USART_Init(USART1, &USART_InitStructure);
        USART_ITConfig(USART1, USART_IT_RXNE, ENABLE);
        USART_Cmd(USART1, ENABLE);

        NVIC_PriorityGroupConfig(NVIC_PriorityGroup_0);
        NVIC_InitTypeDef NVIC_InitStructure;
        NVIC_InitStructure.NVIC_IRQChannel = USART1_IRQn;
        NVIC_InitStructure.NVIC_IRQChannelSubPriority = 0;
        NVIC_InitStructure.NVIC_IRQChannelCmd = ENABLE;
        NVIC_Init(&NVIC_InitStructure);

}


int main(void)
{
        uint8_t stateOld = 3;
        init_UART();
        while(1){
                while(!connected);
                GPIO_SetBits(GPIOA, GPIO_Pin_2);
                while (connected) {
```

```
                    state = GPIO_ReadInputDataBit(GPIOA,
GPIO_Pin_1);
                    if (state != stateOld) {
                            stateOld = state;
                            while(USART_GetFlagStatus(USART1,
USART_FLAG_TXE) == RESET);
                            if (state==0)
USART_SendData(USART1, 'P'); else USART_SendData(USART1,
'N');
                            Delay(100000);
                    }
            }
            GPIO_ResetBits(GPIOA, GPIO_Pin_2);
        }
}


void USART1_IRQHandler(void)
{
        if (USART_GetITStatus(USART1, USART_IT_RXNE) != RESET)
        {
            char data = USART_ReceiveData(USART1);
            if(connected){
                    if (data=='a') GPIO_SetBits(GPIOA,
GPIO_Pin_0);
                    if (data=='b') GPIO_ResetBits(GPIOA,
GPIO_Pin_0);
                    if (data=='r') {
                            while(USART_GetFlagStatus(USART1,
USART_FLAG_TXE) == RESET);
                            if (state==0)
USART_SendData(USART1, 'P'); else USART_SendData(USART1,
'N');
                    }
            }
            if(firstLf) {
                    if(!firstChar) firstChar = data;
            }
            if (data=='\n'){
                    if (firstLf) {
                            firstLf = 0;
                            if (firstChar == 'C') {
                                    connected = 1;
```

```
            }
            if (firstChar == 'D') {
                    connected = 0;
            }
            firstChar = 0;
    } else firstLf = 1;
        }
    }
}
```

6.12. Software UART

Some microcontrollers, for example most from the ATtiny family, don't have any UART. Some have less UARTs than you need. What can you do then? You can create purely software UART. It won't be as good as hardware UART but should be enough for many cases. Let's see what you should keep in mind when you want to send and receive data without help of integrated hardware UART. Although source code for software UART is simple, it would be good if you know some background aspects.

As you know, there are standard speeds of serial transmission. In our examples the speed is 9600 bps but there are many other values, for example 19200 bps or 57600 bps. A very important thing is to keep as close to chosen frequency as possible. If your device is expected to send data at 9600 bps you can't transmit at 10600 bps. That's because the receiver would expect each bit a little bit later than you actually sent it. This way it can miss last bit. If the difference of speed is greater, then more bits can be missed. On the other hand, when you send data to slowly, some transmitted bits can be interpreted as beginning of a next byte. Thus you have to keep your speed as close as possible to the chosen one. If the deviation is about 5% then there is some chance that the communication will be correct but for reliable transmission, especially for higher speeds and longer distances the deviation should be below 1%. Remember that not only your device can work with speed different than required, the other part can introduce some deviation too. Both errors add together. Therefore pay attention to providing appropriate frequency.

However providing an appropriate frequency is not always an easy task. How do we generate frequencies on microcontrollers? Generally we do it by dividing the frequency of the system clock. How is it done? By some mathematical operation? In some sense yes. We use the system clock to increment or decrement value of some register at each tick of this clock. When the register reaches some value we can do something, e.g. toggle state of a pin. This way system clock is divided by some value. This is how timers work. They are circuits which generate interrupts when their counter register overflows or reaches some value. When there is an interrupt we know that some given time elapsed. You can notice that this way the system clock can be divided only by integer values, so you can't get frequency like 2/3 of system clock. OK, but what about standard speeds of serial transmission? Can we achieve them without error? Well, it depends. As it was said, we can divide only by integer numbers. Let's say we have 16 MHz crystal as clock source for our microcontroller and we want to communicate at exactly 9600 bps. Is it possible? No, because 16 000 000 is not multiple of 9600 and cannot by divided integrally by 9600. 16 000 000 / 9600 is 1666.67. The closest integer number is 1667 which differs by about 0.02%. Quite good. But in practice we don't always have timers of appropriate features to perform such operations. Some are only 8-bit and some can't generate an interrupt at arbitrary value. Therefore you need to check out what your microcontroller can offer you. Moreover you need at least three interrupts for each transmitted bit if you want to use software based UART. That's because when you detect start bit you need to be sure that following bits will be sampled somewhere in their middles, not at edges.

In datasheets of many microcontrollers you can find tables with calculated values of appropriate UART register for different speeds of communications and different crystals. You can notice that for some crystals the error is zero for almost all speeds. Such crystals are for example 3.6864 MHz or 11.0592 MHz. Why? Because such frequencies can be integrally divided by popular speeds of serial communication. For example 3 686 400 / 9600 = 384. Thus you should choose one of these crystals for your device if possible. It will make creating software UART a lot easier. It's some kind of paradox, but because of this fact sometimes you can achieve higher speed with crystal of lower frequency than higher. For example for hardware UART of ATmega32 with 16 MHz crystal the error at 115 200 bps is 3.5% which is not acceptable. With 3.6864 MHZ crystal you can operate at this speed because error is zero.

We mentioned timers because they provide a way of performing operations at precisely defined moments in time. But the sole fact that you use a timer doesn't mean that your code will keep good timing. For example, if you wait for some number of timer's interrupts to introduce a delay, its length in the end is going to be a little bit longer because code responsible for checking the number of elapsed interrupts also needs some time to execute. Another tricky situation is when you have a timer which generates interrupt on overflow but you don't want the timer to count from zero. You would like it to count from other value so less increments are required for an overflow and then overflows happen more often. To achieve it you load the value to timer's counter at each interrupt so it won't be counting from zero. But this reload takes time and introduces some delay. Therefore you can observe that your microcontroller generates time intervals a little bit different than you have calculated. A very good illustration to this problem you can find in official Atmel's example: "AVR304: Half Duplex Interrupt Driven Software UART". The source code is in assembler but is well documented so it's not hard to analyze. In the handler for INT0 interrupt a new value is written to Timer0's counter. As you can see the value is calculated using information about number of clock cycles needed to execute code for that handler. In other words, the value is corrected by the delay introduced by interrupt service routine. To know this delay you have to know what instructions are executed by a microcontroller and how long each one of them takes. If you write in assembly language it's quite easy task because assembler instructions are directly translated by a compiler into native code. If you write in C you need to examine machine (native) code generated by your compiler.

Let's sum up information needed for writing software UART:

- choose crystal of frequency integrally dividable by speed of transmission you want to use

- you need at least three interrupts for each bit

- you need a timer which can generate interrupt at arbitrary value

- set and read bits in timer's interrupt routine

6.12.1.   AVR code

Presented below is an example implementation of software UART for ATmega32. It uses Timer1 which is a 16-bit timer which gives good range of allowed values of its counter. The other important thing is that it can generate interrupts on compare (when bit *OCIE1A* is set in *TIMSK* register). The counter is constantly compared with value stored in *OCR1A* register. When they match then an interrupt is generated in next clock cycle. There is no clock division for timer's counter so it is incremented with every clock cycle (bit *CS10* is set in *TCCR1B* register). Counter is cleared on match so it restarts from zero (bit *WGM12* in *TCCR1B* register). The value stored in *OCR1A* is calculated by compiler, basing on crystal frequency defined by *F_CPU* and baud rate defined by *BAUDRATE*. The result of division is further divided by three because the code waits for three periods during each bit. The code can theoretically work with two periods per bit but because of signal fall and rise times and noise it won't be reliable. At the end 1 is subtracted because interrupt doesn't occur immediately but in next clock cycle. It would be good if you also do the calculation by yourself so you can estimate the error.

The main code is inside Timer1 compare interrupt. It consists of several parts responsible for different states of software UART. This is full-duplex solution and sending and receiving are performed independently, there are separate variables for processing incoming and outgoing data. The interrupt routine uses states to distinguish between particular moments in transmission.

Sending is started with *softwareUartSend()* function. Its purpose is to send a single byte but its parameter is 16-bit. That's because of an idea to put also start and stop bit into a variable. It will be explained in a moment. On the beginning the function waits for previous transmission to finish. When the transmitter is in idle state then *cyclesCounterTx* is set to 1. This variable counts clock cycles and is used to introduce delays. If it is set to 1 the transmission will start immediately. Next step is to set counter of bits to transmit. We set it to 10 because there are 10 bits transmitted: start bit, eight bits of data and stop bit. And now, as it was mentioned a moment before, we put all bits together. First we shift the byte we want to send by one position left. This way we get zero on right side as least significant bit. The next operation done in the same line of code is logical OR which sets tenth bit to 1. Because bits are sent from the least significant to the most significant, the transmission starts with start bit we obtained by bit shift, then there are 8 data bits and in the end the stop bit we got from logical OR. When we have everything set up we change state of transmitter so timer's interrupt routine can do rest of the job.

If the state of transmitter is set to *TRANSMIT* then the routine waits for appropriate moment to output transmitted bit on microcontroller's pin. Waiting is performed by observing cycles counter. This counter is decremented on every interrupt in *TRANSMIT* state. When the counter is zero then the pin is set to one or zero depending on the least significant bit of the variable which stores bits to send. Then the variable is shifted right so next time next bit can be checked. When all bits are sent the transmitter changes state to *TRANSMIT_STOP*. This additional state is needed because we can't start sending a new byte until the stop bit is not finished.

When the receiver is idle, it checks state of the pin chosen as receiver's input pin. If it is zero it means we have a start bit. Then we set *cyclesCounterRx* to 4 so the first bit of data will be sampled somewhere in the middle, between 1/3 and 2/3 of its period. Sampling is done when receiver's state is set to *RECEIVE*. The variable storing already received bits is shifted right to make space for incoming bit. This new bit is written as the most significant one. As you remember bits are transmitted starting from the least significant one. With shifting they are at their right positions after 8 shifts. The first one will be moved to the least significant position and following bits will be placed more and more left (to the more significant positions) with last bit at most significant position. When all bits are received the receiver goes into *RECEIVE_STOP* state. We need it because otherwise last data bit could be interpreted as start bit of next transmitted byte.

Fact of receiving a byte is signaled by setting *softwareUartNewData* to 1. You can check this variable in your program to detect that a new byte is available in *softwareUartInputData*. But you may want to process incoming data immediately during the timer's interrupt. As you can see in the code, there is a commented out call to *processIncomingData()*. It's a hint that you can create a function for handling incoming bytes, for example storing them in a buffer. It will be called during an interrupt so remember that it should be short so it can finish before next interrupt is generated. And if you are going to use some global variables there then don't forget to declare them as *volatile*.

Presented software base UART doesn't support parity checking or flow control. One stop bit and eight bits of data transmission mode is supported only. This is however enough for most cases. If you want more features, you have to adapt the code. For changing number of data or stop bits you need just to change appropriate numbers. In case of parity checking and flow control some extra logic must be added.

Serial port and microcontrollers

As this is a demonstration of the software UATRT itself, the *main()* loop doesn't do anything specific. It just sends back every received byte incremented by one.

```c
#include <avr/interrupt.h>
#include <avr/io.h>

#define F_CPU 16000000UL
#define BAUDRATE 9600UL

enum SoftwareUartStateRx {IDLE_RX, RECEIVE, RECEIVE_STOP};
enum SoftwareUartStateTx {IDLE_TX, TRANSMIT, TRANSMIT_STOP};
volatile enum SoftwareUartStateRx softwareUartStateRx =
IDLE_RX;
volatile enum SoftwareUartStateTx softwareUartStateTx =
IDLE_TX;
volatile uint8_t cyclesCounterRx, cyclesCounterTx;
volatile uint8_t softwareUartBitsToSend;
volatile uint16_t softwareUartOutputData;
volatile uint8_t softwareUartInputData;
volatile uint8_t softwareUartNewData = 0;

#define PORT_RX PIND //RX port
#define PIN_RX PIND2 //pin of RX port
#define PORT_TX PORTD //TX port
#define PIN_TX PORTD3 //pin of TX port
#define DDR_TX DDRD //data direction register for TX port

void init_Timer() {
      TIMSK = _BV(OCIE1A); //interrupt on compare
      TCCR1B = _BV(WGM12)|_BV(CS10); //no clock division
      OCR1A = F_CPU / BAUDRATE / 3 - 1; //maximum value for
timer's counter
}

ISR(TIMER1_COMPA_vect)
{
      static uint8_t bitsToReceive;
      static uint8_t inData;

      if (softwareUartStateTx == TRANSMIT) {
            if (--cyclesCounterTx == 0) {
```

```
                    if(softwareUartOutputData & 1) PORT_TX |=
_BV(PIN_TX); else PORT_TX &= ~_BV(PIN_TX); //data bit
                    softwareUartOutputData >>= 1;
                    cyclesCounterTx = 3;
                    if (--softwareUartBitsToSend == 0) {
                            softwareUartStateTx =
TRANSMIT_STOP;
                    }
                }
        }

        if (softwareUartStateTx == TRANSMIT_STOP) {
                if (--cyclesCounterTx == 0) {
                        softwareUartStateTx = IDLE_TX;
                }
        }

        if ((softwareUartStateRx == IDLE_RX) && !(PORT_RX &
_BV(PIN_RX))) {
                cyclesCounterRx = 4;
                softwareUartStateRx = RECEIVE;
                bitsToReceive = 8;
                inData = 0;
        }

        if (softwareUartStateRx == RECEIVE) {
                if (--cyclesCounterRx == 0) {
                        inData >>= 1;
                        if (PORT_RX & _BV(PIN_RX)) inData |= 0x80;
                        cyclesCounterRx = 3;
                        if (--bitsToReceive == 0) {
                                softwareUartStateRx = RECEIVE_STOP;
                                softwareUartInputData = inData;
                                softwareUartNewData = 1;
                                //processIncomingData();
                        }
                }
        }

        if (softwareUartStateRx == RECEIVE_STOP) {
                if (--cyclesCounterRx == 0) {
                        softwareUartStateRx = IDLE_RX;
                }
```

```
        }
}

void softwareUartSend(uint16_t data) {
        while(softwareUartStateTx != IDLE_TX);
        cyclesCounterTx = 1;
        softwareUartBitsToSend = 10;
        softwareUartOutputData = (data << 1) | 0b1000000000;
        softwareUartStateTx = TRANSMIT;
}

int main(void)
{
        init_Timer();
        sei();
        DDR_TX = _BV(PIN_TX);
        PORT_TX |= _BV(PIN_TX);
        while(1)
        {
                if(softwareUartNewData) {
                        softwareUartNewData = 0;
                        softwareUartSend(softwareUartInputData +
1);
                }
        }
}
```

6.12.2.   STM32 code

The approach for STM32 is very similar to AVR version and there is not much to talk about. You just have to remember that *TIM_Period* field is 16-bit so it can't hold values greater than 65 535. This may happen when you have high clock frequency and very low speed of transmission, like 300 bps. Then you need to use prescaler.

```
#include "stm32f10x.h"

enum SoftwareUartStateRx {IDLE_RX, RECEIVE, RECEIVE_STOP};
enum SoftwareUartStateTx {IDLE_TX, TRANSMIT, TRANSMIT_STOP};
volatile enum SoftwareUartStateRx softwareUartStateRx =
IDLE_RX;
```

```
volatile enum SoftwareUartStateTx softwareUartStateTx =
IDLE_TX;
volatile uint8_t cyclesCounterRx, cyclesCounterTx;
volatile uint8_t softwareUartBitsToSend;
volatile uint16_t softwareUartOutputData;
volatile uint8_t softwareUartInputData;
volatile uint8_t softwareUartNewData = 0;

#define PORT_RX GPIOA //RX port
#define PIN_RX GPIO_Pin_0 //pin of RX port
#define PORT_TX GPIOA //TX port
#define PIN_TX GPIO_Pin_1 //pin of TX port
#define RCC_TX RCC_APB2Periph_GPIOA
#define RCC_RX RCC_APB2Periph_GPIOA

#define F_CPU 72000000UL
#define BAUDRATE 9600UL

void init_GPIO();
void init_Timer();

void init_Timer() {
        RCC_APB1PeriphClockCmd(RCC_APB1Periph_TIM2, ENABLE);

        TIM_TimeBaseInitTypeDef TIM_InitStruct;
        TIM_InitStruct.TIM_ClockDivision = TIM_CKD_DIV1;
        TIM_InitStruct.TIM_CounterMode = TIM_CounterMode_Up;
        TIM_InitStruct.TIM_Period = F_CPU / BAUDRATE / 3 - 1;
        TIM_InitStruct.TIM_Prescaler = 0;
        TIM_TimeBaseInit(TIM2, &TIM_InitStruct);

        TIM_ClearFlag(TIM2, TIM_FLAG_Update);
        TIM_ITConfig(TIM2, TIM_IT_Update, ENABLE);
        TIM_Cmd(TIM2, ENABLE);

        NVIC_InitTypeDef NVIC_InitStruct;
        NVIC_InitStruct.NVIC_IRQChannel = TIM2_IRQn;
        NVIC_InitStruct.NVIC_IRQChannelCmd = ENABLE;
        NVIC_InitStruct.NVIC_IRQChannelPreemptionPriority = 0;
        NVIC_InitStruct.NVIC_IRQChannelSubPriority = 0;
        NVIC_Init(&NVIC_InitStruct);

}
```

```
void init_GPIO() {
      RCC_APB2PeriphClockCmd(RCC_RX | RCC_TX, ENABLE);

      GPIO_InitTypeDef GPIO_InitStructure;
      GPIO_InitStructure.GPIO_Pin = PIN_RX;
      GPIO_InitStructure.GPIO_Speed = GPIO_Speed_2MHz;
      GPIO_InitStructure.GPIO_Mode = GPIO_Mode_IN_FLOATING;
      GPIO_Init(PORT_RX, &GPIO_InitStructure);

      GPIO_InitStructure.GPIO_Pin = PIN_TX;
      GPIO_InitStructure.GPIO_Mode = GPIO_Mode_Out_PP;
      GPIO_Init(PORT_TX, &GPIO_InitStructure);

      GPIO_SetBits(PORT_TX, PIN_TX);
}

void TIM2_IRQHandler() {
      if (TIM_GetITStatus(TIM2, TIM_IT_Update)!=RESET) {
            TIM_ClearFlag(TIM2, TIM_FLAG_Update);

            static uint8_t bitsToReceive;
            static uint8_t inData;

            if (softwareUartStateTx == TRANSMIT) {
                  if (--cyclesCounterTx == 0) {
                        if(softwareUartOutputData & 1)
GPIO_SetBits(PORT_TX, PIN_TX); else GPIO_ResetBits(PORT_TX,
PIN_TX); //data bit
                        softwareUartOutputData >>= 1;
                        cyclesCounterTx = 3;
                        if (--softwareUartBitsToSend == 0)
{
                              softwareUartStateTx =
TRANSMIT_STOP;
                        }
                  }
            }

            if (softwareUartStateTx == TRANSMIT_STOP) {
                  if (--cyclesCounterTx == 0) {
                        softwareUartStateTx = IDLE_TX;
                  }
```

```
            }

            if ((softwareUartStateRx == IDLE_RX) &&
!(GPIO_ReadInputDataBit(PORT_RX, PIN_RX))) {
                    cyclesCounterRx = 4;
                    softwareUartStateRx = RECEIVE;
                    bitsToReceive = 8;
                    inData = 0;
            }

            if (softwareUartStateRx == RECEIVE) {
                    if (--cyclesCounterRx == 0) {
                            inData >>= 1;
                            if (GPIO_ReadInputDataBit(PORT_RX,
PIN_RX)) inData |= 0x80;
                            cyclesCounterRx = 3;
                            if (--bitsToReceive == 0) {
                                    softwareUartStateRx =
RECEIVE_STOP;
                                    softwareUartInputData =
inData;
                                    softwareUartNewData = 1;
                                    //processIncomingData();
                            }
                    }
            }

            if (softwareUartStateRx == RECEIVE_STOP) {
                    if (--cyclesCounterRx == 0) {
                            softwareUartStateRx = IDLE_RX;
                    }
            }

    }
}

void softwareUartSend(uint16_t data) {
      while(softwareUartStateTx != IDLE_TX);
      cyclesCounterTx = 1;
      softwareUartBitsToSend = 10;
      softwareUartOutputData = (data << 1) | 0x200;
      softwareUartStateTx = TRANSMIT;
}
```

```
int main(void)
{
        init_Timer();
        init_GPIO();

        while(1)
        {
                if(softwareUartNewData) {
                        softwareUartNewData = 0;
                        softwareUartSend(softwareUartInputData +
1);
                }
        }
}
```

6.13.   Going deeper: DMA

When serial data are coming to a microcontroller, we have generally two choices: check for new data in main loop by constantly polling appropriate register, or catch the data in an interrupt handler. In both those cases we need to have a code responsible for handling each incoming byte. But some microcontrollers offer third way: DMA, which stands for Direct Memory Access. With DMA data coming from a peripheral can be automatically transmitted to specified place in memory without any help from the core of the microcontroller. The other way it works analogically. From the point of view of your program you don't have to deal with a peripheral when you want to read incoming data or send some information, you just operate on buffers in memory. It is a good solution especially for large packets of data, transmitting with high speed. It is also possible to pass data directly between peripherals, for example between a USART and a timer.

DMA is a feature of more advanced microcontrollers and is not available on ATmega32. However it's very common in STM32 family, including STM32F103CBT6. Let's see how DMA works on this chip.

6.13.1.   Hardware configuration

Because we are going just to process data internally in a microcontroller, we don't need any peripheral elements like LEDs, buttons, etc. There is just standard circuit for power supply, quartz and RS-232 voltage converter.

6.13.2.   STM32 code

The first DMA example shows how to process data received with DMA transfer and how to send data with DMA. Presented code calculates CRC32 (cyclic redundancy check) sum for received data. Of course you can compute CRC32 sums on your computer, it will be faster and simpler. But calculating CRC32 on a microcontroller fits well for the subject of data processing so it can serve as a good illustration. Moreover calculating of checksums is often performed on microcontrollers to detect errors in data, caused by faulty media (various kinds of memories) or noisy transmission channels.

Before we start to use DMA on STM32 chips we have to know, that DMA transfers are performed for some predefined amount of data. When all bytes have been transmitted then a flag is set and/or an interrupt is generated. Therefore we need to transmit data as packets of the same size, which somewhat complicates both preparing data to be sent and analyzing received data.

Let's look at the source code.

```
#include "stm32f10x.h"

#define DMA_BLOCK_LENGTH 256

void init_USART();
void init_DMA();
void startUSARTDMATransferIx();
uint32_t calcCRC32HW(uint32_t * buf, u32 size32);
uint32_t calcCRC32SW(u8 * buf, u32 size, uint32_t
startValue);
u32 revbit(u32 data);

uint8_t dataBuffer[DMA_BLOCK_LENGTH];
uint32_t crcValue;
uint8_t firstBlock = 1;
uint32_t dataLength = 0;
uint32_t * dataBufferPtr;

uint32_t calcCRC32HW(uint32_t * buf, u32 size32)
{
        while(size32--) CRC->DR = revbit(*(buf++));
        return revbit(CRC->DR);
}
```

```
uint32_t calcCRC32SW(u8 * buf, uint32_t size, uint32_t
currentValue){
      while(size--)
      {
            currentValue ^= (u32)*buf++;
            for(uint32_t j=0; j<8; j++)
                  if (currentValue & 1)
                      currentValue = (currentValue >> 1) ^
0xEDB88320;
                  else
                      currentValue >>= 1;
      }
      return ~currentValue;
};

int main()
{
      RCC_AHBPeriphClockCmd(RCC_AHBPeriph_CRC, ENABLE);
      init_USART();
      init_DMA();

      while(1) {

      }
}

void DMA1_Channel5_IRQHandler(void)
{
   DMA_ClearITPendingBit(DMA1_IT_TC5);
   if(firstBlock){
            CRC_ResetDR();
            dataLength = *((uint32_t *) dataBuffer);
            dataBufferPtr = (uint32_t *) dataBuffer;
            dataBufferPtr++;
      }
      if(dataLength <= (DMA_BLOCK_LENGTH + (firstBlock ? -4 :
0))){ //last block
            crcValue = calcCRC32HW(dataBufferPtr, dataLength
/ 4);
            dataBufferPtr += dataLength / 4;
            crcValue = calcCRC32SW((uint8_t *)dataBufferPtr,
dataLength % 4, crcValue);
            startUSARTDMATransferTx();
```

```
                firstBlock = 1;
        } else { //not last block
                uint32_t processedDataLength = DMA_BLOCK_LENGTH
/ 4 + (firstBlock ? -1 : 0);
                calcCRC32HW(dataBufferPtr, processedDataLength);
                dataBufferPtr = (uint32_t *) dataBuffer;
                dataLength -= processedDataLength * 4;
                firstBlock = 0;
        }
}


u32 revbit(u32 data)
{
        asm("rbit r0,r0");
        return data;
};

void startUSARTDMATransferTx() {
        DMA_Cmd(DMA1_Channel4, DISABLE);
        DMA1_Channel4->CMAR = (uint32_t)&crcValue;
        DMA1_Channel4->CNDTR = 4;
        DMA_Cmd(DMA1_Channel4, ENABLE);
}

void init_USART(){

        RCC_APB2PeriphClockCmd(RCC_APB2Periph_GPIOA |
RCC_APB2Periph_USART1, ENABLE);

        GPIO_InitTypeDef GPIO_InitStructure;
        GPIO_InitStructure.GPIO_Pin = GPIO_Pin_9;
        GPIO_InitStructure.GPIO_Speed = GPIO_Speed_2MHz;
        GPIO_InitStructure.GPIO_Mode = GPIO_Mode_AF_PP;
        GPIO_Init(GPIOA, &GPIO_InitStructure);

        GPIO_InitStructure.GPIO_Pin = GPIO_Pin_10;
        GPIO_InitStructure.GPIO_Mode = GPIO_Mode_IN_FLOATING;
        GPIO_Init(GPIOA, &GPIO_InitStructure);

        USART_InitTypeDef USART_InitStructure;
        USART_InitStructure.USART_BaudRate = 115200;
```

```
        USART_InitStructure.USART_WordLength =
USART_WordLength_8b;
        USART_InitStructure.USART_StopBits = USART_StopBits_1;
        USART_InitStructure.USART_Parity = USART_Parity_No;
        USART_InitStructure.USART_HardwareFlowControl =
USART_HardwareFlowControl_None;
        USART_InitStructure.USART_Mode = USART_Mode_Tx |
USART_Mode_Rx;
        USART_Init(USART1, &USART_InitStructure);
}

void init_DMA(){
        RCC_AHBPeriphClockCmd(RCC_AHBPeriph_DMA1, ENABLE);

        DMA_InitTypeDef DMA_InitStructure;

        DMA_DeInit(DMA1_Channel5);
        DMA_InitStructure.DMA_PeripheralBaseAddr =
(uint32_t)&(USART1->DR);
        DMA_InitStructure.DMA_MemoryBaseAddr = (uint32_t)
dataBuffer;
        DMA_InitStructure.DMA_DIR = DMA_DIR_PeripheralSRC;
        DMA_InitStructure.DMA_BufferSize = DMA_BLOCK_LENGTH;
        DMA_InitStructure.DMA_PeripheralInc =
DMA_PeripheralInc_Disable;
        DMA_InitStructure.DMA_MemoryInc = DMA_MemoryInc_Enable;
        DMA_InitStructure.DMA_MemoryDataSize =
DMA_MemoryDataSize_Byte;
        DMA_InitStructure.DMA_PeripheralDataSize =
DMA_PeripheralDataSize_Byte;
        DMA_InitStructure.DMA_Mode = DMA_Mode_Circular;
        DMA_InitStructure.DMA_Priority = DMA_Priority_High;
        DMA_InitStructure.DMA_M2M = DMA_M2M_Disable;
        DMA_Init(DMA1_Channel5, &DMA_InitStructure);

        DMA_DeInit(DMA1_Channel4);
        DMA_InitStructure.DMA_MemoryBaseAddr =
(uint32_t)&crcValue;
        DMA_InitStructure.DMA_DIR = DMA_DIR_PeripheralDST;
        DMA_InitStructure.DMA_BufferSize = 4;
        DMA_InitStructure.DMA_Mode = DMA_Mode_Normal;
        DMA_Init(DMA1_Channel4, &DMA_InitStructure);
```

```
    NVIC_InitTypeDef NVIC_InitStructure;
    NVIC_PriorityGroupConfig(NVIC_PriorityGroup_0);
    NVIC_InitStructure.NVIC_IRQChannel =
DMA1_Channel5_IRQn;
    NVIC_InitStructure.NVIC_IRQChannelPreemptionPriority =
0;
    NVIC_InitStructure.NVIC_IRQChannelSubPriority = 1;
    NVIC_InitStructure.NVIC_IRQChannelCmd = ENABLE;
    NVIC_Init(&NVIC_InitStructure);

    DMA_ITConfig(DMA1_Channel5, DMA_IT_TC, ENABLE);
    USART_DMACmd(USART1, USART_DMAReq_Rx | USART_DMAReq_Tx,
ENABLE);
    DMA_Cmd(DMA1_Channel5, ENABLE);
    USART_Cmd(USART1, ENABLE);
}
```

In our example we configure two DMA channels: channel 5 for transmitting data from USART1 to memory and channel 4 for the other way. Why channels 4 and 5? They are the DMA channels which are assigned to USART1 in the chip. If you are interested in mappings for other channels, refer to *Table 65. Summary of DMA1 requests for each channel* in *STM32 Reference manual* (document *RM0008*).

Almost all DMA configuration is done with *DMA_Init()* function and its associated structure. Although channel 5 works with USART1 we have to provide address of USART1's data register. We do it with *DMA_PeripheralBaseAddr* field of configuration structure. Analogically with *DMA_MemoryBaseAddr* field we set place in memory where the received data will be written. The actual direction of data flow we set using *DMA_DIR* field. Here the peripheral (USART1) works as the source of data. *DMA_BufferSize* is an important field, it defines size of DMA transfer. Here the size is 256 bytes, the same as size of the buffer where we are going to store received data. During the DMA transfer of incoming data all bytes are read from the same place, USART1's data register, but they go to consecutive places in memory. Therefore we don't increment peripheral address but do increment for memory. Data are transmitted as bytes in our example and they are stored as bytes so we set data size to byte for both memory and peripheral. The field *DMA_Mode* is set to *DMA_Mode_Circular* because we need the channel to work in circular mode. This is because after finishing one transfer we want the next one to start at the beginning of the buffer in memory. Hence data from each transfer are saved at the same place, overwriting data from previous transfer. Priority of the DMA channel is set to high, memory-to-memory transfers are disabled.

Channel 4 is configured in the same way, using the same structure, just some fields are modified. Because we use this channel to send data we have to modify memory address to point to the variable we are going to send. We also change the direction, now the peripheral is the destination. The CRC32 sum is, as the name says, a 32-bit (4 byte) value, therefore we set *DMA_BufferSize* to 4. For sending we don't want to work in circular mode. When the transfer is finished we don't want it to automatically start over. That why we use normal mode.

We would like to know that a DMA transfer has finished and a new packet is ready to process. We have two choices here: we can check state of *DMA1_FLAG_TC5* flag or use an interrupt. In the example we use the interrupt so the lines responsible for the flag are commented out. The interrupt is configured in a standard way, using *NVIC_Init()* function and appropriate structure. Then we enable the interrupt for channel 5, enable DMA transfer for USART1's transmitter and receiver, enable DMA channel 5 and finally enable the USART. USART1 is configured in a usual way, but the speed is set to 115 200 bits per second so you can wait shorter for transmitting larger files.

As you have probably noticed, we haven't enabled DMA channel 4. That's because we don't want to send the *crcValue* variable when the program starts but when the CRC32 sum is ready. To start DMA transmission we need to disable the channel, set pointer to the place where the variable we want to send is located, set size of transfer and then enable the channel. In our example it is realized using *startUSARTDMATransferTx()* function. We operate here directly on DMA channel registers. We could use *DMA_Init()* function from the STM32 Standard Peripheral Library and fill the structure but it would be an overkill.

OK, let's now talk about calculating CRC32 sum. STM32 microcontrollers have built-in CRC sum generator. It is very simple, you just need to operate on its two registers: *DR* and *CR*. The first one is a data register after writing a 32-bit value the generator starts to process it and stores the result in the same register. Hence all you have to do is to write all the data you want to get CRC sum for and then read the result. The second register is a control register. Writing 1 to this register causes reset of the CRC generator. In our code we call *CRC_ResetDR()* function which performs this operation. You can replace this call with simple write:

```
CRC->CR = 1;
```

The STM32 CRC generator is very simple but there are two problems with it: it works with 32-bit values and it generates CRC sums different than in most popular CRC32 implementation (e.g. the one you can try at http://crc32-checksum.waraxe.us/). To solve the first problem we have to calculate CRC32 for remaining 1 to 3 bytes by software. For example if we have 10 bytes of data and we want to calculate CRC32 sum for them we need to write first four bytes to the *DR* register, then second four bytes again to this register, take the result from the register and the remaining two bytes and calculate the sum by software.

As for the second problem we can solve it using bit reversing. Every time we write to *DR* register we have to invert order of bits in a 32-bit value we are going to write. We have to do the same when we are reading the result. And finally we need to negate all the bits in the final result.

Let's look at functions responsible for calculating the CRC32 sum. *calcCRC32HW()* uses hardware CRC generator. Its's simple loop which writes contents of given buffer to the *DR* register. It writes 4 bytes at a time. The *size32* tells how many such writes to perform, in other words it is size of input data expressed as number of quadruples. The result is taken from the *DR* register. For both reading and writing *DR* we reverse order of bits. It is done using revbit() function which is just a call to single microcontroller's operation: rbit. This call is realized by a piece of assembler code inserted into C code. Here we use the fact that argument of the function (*data*) is placed in *r0* register of our chip. The *rbit* operation reverses bits in this register and stores the result in the same register. This way the *data* parameter is modified. Finally it is returned as the result of the function. The *calcCRC32SW()* function calculates CRC32 by software and is designed to finish calculations started by *calcCRC32HW()*. It takes value obtained from that function, performs calculations for remaining bytes, negates bits in calculated sum and returns the result.

Now we are going to analyze main logic of the program. At the beginning we enable clock for hardware CRC generator and configure USART and DMA. Then the program enters an infinite loop and all operations are performed when transfer complete interrupt from DMA channel 5 occurs. Handler for this interrupt is responsible for processing packet of data which just arrived.

It has been assumed that the first thing sent to the microcontroller is total length of data to process. The length is encoded using first four bytes of the first packet. Therefore in the first packet we transmit less than can be transmitted in following packets. Let's say we want to compute CRC32 sum for data 600 bytes in size and the size of single packet is 256 bytes. Hence the first packet would consist of 4 bytes of length and 252 bytes of data. The second one would contain 256 bytes of data and the third one 92 bytes of data. The remaining 164 bytes in the third byte can have arbitrary values and they don't affect anything.

When the first packet (block of data) arrives, we read data length from its first four bytes and initialize pointer to point at the beginning of data. The pointer is for 32-bit values so when we increment it, it jumps from the first received byte to the fifth one (from byte with index 0 to byte with index 4). Then we check if the packet we received is the last one or not. We can know it by checking the *dataLength* variable which holds number of bytes to process. If it is the last packet we perform hardware CRC computation and then software CRC computation. The result is sent using DMA transfer. If the current packet is not the last one we perform just hardware computation. The buffer pointer is set to point at the beginning of the buffer and *dataLength* is decremented by number of bytes which have been processed. Finally we set *firstBlock* to zero to be sure that with the next packet we won't skip first four bytes as just only the first packet contains length at its beginning.

I you want to wait for the DMA transfer to complete you can check the TC flag:

e.g. `while(DMA_GetFlagStatus(DMA1_FLAG_TC5)!=RESET);`

The flag can be cleared like this: *DMA_ClearFlag(DMA1_FLAG_TC5);*

That's pretty everything. Now you know how to use DMA to transfer data between memory and a peripheral. As you can see there is no code we used in previous examples for sending and receiving data with serial port. We have also said a few words about some basic aspects of data processing and packet transmission.

6.13.3.   PC code

For our SMT32 code we need also some PC application so we can send data to our microcontroller for computation and read received values. Here is the code.

```csharp
using System;
using System.Windows.Forms;
using System.IO.Ports;
using System.Text;
using System.IO;

namespace SerialSample13
{
    public partial class Form1 : Form
    {
        SerialPort ComPort = new SerialPort("COM1");
        const int DMA_BLOCK_LENGTH = 256;
        string filePath = "";
        uint crcSum;
        public Form1()
        {
            InitializeComponent();
            ComPort.BaudRate = 115200;
            ComPort.Parity = Parity.None;
            ComPort.StopBits = StopBits.One;
            ComPort.DataBits = 8;
            ComPort.Handshake = Handshake.None;
            ComPort.DataReceived += OnSerialDataReceived;
            ComPort.Open();
        }

        private void OnSerialDataReceived(object sender,
SerialDataReceivedEventArgs args)
        {
            crcSum = 0;
            for (int i = 0; i < 4; i++)
            {
                crcSum >>= 8;
                crcSum |= (uint)(ComPort.ReadByte() << 24);
            }
            lblCRC.Invoke(new
EventHandler(lblCRC_DataReceived));
```

```
        }

        private void lblCRC_DataReceived(object sender,
EventArgs e)
        {
            lblCRC.Text = crcSum.ToString("X8");
        }

        private void btnSend_Click(object sender, EventArgs
e)
        {
            byte[] buffer = new byte[DMA_BLOCK_LENGTH];
            byte[] strBytes =
Encoding.ASCII.GetBytes(textBox1.Text);
            byte[] lengthBytes =
BitConverter.GetBytes(strBytes.Length);
            for (int i = 0; i < 4; i++) buffer[i] =
lengthBytes[i];
            Array.Copy(strBytes, 0, buffer, 4,
strBytes.Length);
            ComPort.Write(buffer, 0, DMA_BLOCK_LENGTH);
        }

        private void btnSendFile_Click(object sender,
EventArgs e)
        {
            sendFile();
        }

        private void sendFile()
        {
            if (filePath == "") return;
            byte[] buffer = new byte[DMA_BLOCK_LENGTH];
            using (BinaryReader b = new
BinaryReader(File.Open(filePath, FileMode.Open)))
            {
                int fileLength = (int)b.BaseStream.Length;
                for (int i = 0; i < fileLength; )
                {
                    int readBytes;
                    if (i == 0)
                    {
```

```
                    byte[] lengthBytes =
BitConverter.GetBytes(fileLength);
                    for (int j = 0; j < 4; j++) buffer[j]
= lengthBytes[j];
                    readBytes = b.Read(buffer, 4,
DMA_BLOCK_LENGTH - 4);
                    ComPort.Write(buffer, 0,
DMA_BLOCK_LENGTH);
                }
                else
                {
                    readBytes = b.Read(buffer, 0,
DMA_BLOCK_LENGTH);
                    ComPort.Write(buffer, 0,
DMA_BLOCK_LENGTH);
                }
                i += readBytes;
                progressBar1.Value = i * 100 /
fileLength;
            }
        }
    }

    private void btnBrowse_Click(object sender, EventArgs
e)
    {
        OpenFileDialog openFileDialog = new
OpenFileDialog();
        DialogResult result =
openFileDialog.ShowDialog();
        if (result == DialogResult.OK)
        {
            filePath = openFileDialog.FileName;
            lblPath.Text = "path: " + filePath;
        }

    }

  }
}
```

The program can send data for computation from two sources: a text box or a file. In case of the text box the situation is simple. The typed text is converted into array of bytes, the same happens to length of that text. Then both arrays are copied to a buffer and the buffer is sent through a serial port. When the microcontroller sends back a result it is received byte by byte. Each received byte is places on most significant place in *crcSum* variable. Before each write content of the variable is shifted so room for new byte is made. When all four bytes has been received the value is displayed in hexadecimal format.

For files the algorithm is a little more complicated. We open a file and send its content in a loop. In each iteration we try to read as many bytes as the size of packet. The first packet is an exception because the number of bytes we try to read is decremented by 4. That's of course because of the length sent in first packet. Here we use the nice fact that the *Read()* method returns number of read bytes. We store this value in *readBytes* variable. We use it to increment the *i* counter. As you see this variable isn't incremented always by the same value, that's why there is an empty space after second semicolon in the *for* loop header.

When a block is sent we calculate value for the progress bar. Quick question: isn't the form

```
progressBar1.Value = i / fileLength * 100;
```

more natural than:

```
progressBar1.Value = i * 100 / fileLength;
```

The answer is yes. At least for me. So why the less elegant form has been chosen? Remember that all the values here are integer and hence there are integer mathematical operations executed. Let's say we transmitted 100 bytes of 200 byte file. In the first case the first operation is division, integer division. Because of this 100/200 isn't 0.5 but 0. If we use the first form then the progress bar would be constantly at zero and immediately jump to maximum at the end. Hence we begin with multiplication and then do division. Nonetheless you can begin with division but you need to get rid of integer operations. How? Just cast first variable to double:

```
progressBar1.Value = (int)((double) i / fileLength * 100);
```

We also need to cast the result to *int* because *Value* is of this type.

For choosing files we use standard *OpenFileDialog* class.

6.13.4.   ARM code – second example

With DMA we can also transfer data between peripherals. It's possible because they use registers which are also visible as places in memory. In this example we are going to reimplement driving LED with PWM. The code practically consists of initializations only.

```
#include "stm32f10x.h"

void init_USART();
void init_Timer();
void init_DMA();

int main(void)
{
        init_USART();
        init_Timer();
        init_DMA();
        while(1);
}

void init_USART(){

        RCC_APB2PeriphClockCmd(RCC_APB2Periph_GPIOA |
RCC_APB2Periph_USART1, ENABLE);

        GPIO_InitTypeDef GPIO_InitStructure;
        GPIO_InitStructure.GPIO_Pin = GPIO_Pin_9;
        GPIO_InitStructure.GPIO_Speed = GPIO_Speed_2MHz;
```

```
        GPIO_InitStructure.GPIO_Mode = GPIO_Mode_AF_PP;
        GPIO_Init(GPIOA, &GPIO_InitStructure);

        GPIO_InitStructure.GPIO_Pin = GPIO_Pin_10;
        GPIO_InitStructure.GPIO_Speed = GPIO_Speed_2MHz;
        GPIO_InitStructure.GPIO_Mode = GPIO_Mode_IN_FLOATING;
        GPIO_Init(GPIOA, &GPIO_InitStructure);

        USART_InitTypeDef USART_InitStructure;
        USART_InitStructure.USART_BaudRate = 9600;
        USART_InitStructure.USART_WordLength =
USART_WordLength_8b;
        USART_InitStructure.USART_StopBits = USART_StopBits_1;
        USART_InitStructure.USART_Parity = USART_Parity_No;
        USART_InitStructure.USART_HardwareFlowControl =
USART_HardwareFlowControl_None;
        USART_InitStructure.USART_Mode = USART_Mode_Tx |
USART_Mode_Rx;
        USART_Init(USART1, &USART_InitStructure);
        USART_Cmd(USART1, ENABLE);
}

void init_Timer(){

        RCC_APB1PeriphClockCmd(RCC_APB1Periph_TIM2, ENABLE);

        TIM_TimeBaseInitTypeDef TIM_InitStruct;
        TIM_InitStruct.TIM_ClockDivision = TIM_CKD_DIV1;
        TIM_InitStruct.TIM_CounterMode = TIM_CounterMode_Up;
        TIM_InitStruct.TIM_Period = 255;
        TIM_InitStruct.TIM_Prescaler = 100;
        TIM_TimeBaseInit(TIM2, &TIM_InitStruct);

        TIM_OCInitTypeDef TIM_OCInitStructure;
        TIM_OCInitStructure.TIM_OCMode = TIM_OCMode_PWM1;
        TIM_OCInitStructure.TIM_OutputState =
TIM_OutputState_Enable;
        TIM_OCInitStructure.TIM_Pulse = 0;
        TIM_OCInitStructure.TIM_OCPolarity =
TIM_OCPolarity_High;
        TIM_OC1Init(TIM2, &TIM_OCInitStructure);

        TIM_Cmd(TIM2, ENABLE);
```

```
        GPIO_InitTypeDef GPIO_InitStructure;
        GPIO_InitStructure.GPIO_Pin = GPIO_Pin_0;
        GPIO_InitStructure.GPIO_Speed = GPIO_Speed_2MHz;
        GPIO_InitStructure.GPIO_Mode = GPIO_Mode_AF_PP;
        GPIO_Init(GPIOA, &GPIO_InitStructure);
}

void init_DMA() {
        RCC_AHBPeriphClockCmd(RCC_AHBPeriph_DMA1, ENABLE);

        DMA_InitTypeDef DMA_InitStruct;
        DMA_DeInit(DMA1_Channel5);
        DMA_InitStruct.DMA_BufferSize = 1;
        DMA_InitStruct.DMA_DIR = DMA_DIR_PeripheralSRC;
        DMA_InitStruct.DMA_M2M = DMA_M2M_Disable;
        DMA_InitStruct.DMA_MemoryBaseAddr = (uint32_t)&(TIM2-
>CCR1);
        DMA_InitStruct.DMA_MemoryDataSize =
DMA_MemoryDataSize_HalfWord;
        DMA_InitStruct.DMA_MemoryInc = DMA_MemoryInc_Disable;
        DMA_InitStruct.DMA_Mode = DMA_Mode_Circular;
        DMA_InitStruct.DMA_PeripheralBaseAddr =
(uint32_t)&(USART1->DR);
        DMA_InitStruct.DMA_PeripheralDataSize =
DMA_PeripheralDataSize_HalfWord;
        DMA_InitStruct.DMA_PeripheralInc =
DMA_PeripheralInc_Disable;
        DMA_InitStruct.DMA_Priority = DMA_Priority_Medium;

        USART_DMACmd(USART1, USART_DMAReq_Rx, ENABLE);
        DMA_Init(DMA1_Channel5, &DMA_InitStruct);
        DMA_Cmd(DMA1_Channel5, ENABLE);

}
```

Configurations of timer working in PWM mode and of USART have been already discussed. We were also talking about DMA. Here the difference is that as the memory address we provide address of one of timer's registers. This way each time a byte is received the timer is reconfigured. There are no interrupts and the infinite loop in *main()* doesn't do anything.

You might wonder why data size has been set to half word (16 bits) while we receive single bytes from USART. It turns out timer registers can be accessed by half-words (16-bit) or words (32-bit), not by bytes. If a peripheral doesn't support byte or halfword operations, the data are duplicated on the unused lanes of the HWDATA[31:0] bus. So if for example USART receives 12h byte then we get 12121212h value. That's why we use half word as data size. You can make an experiment and change the size to byte. You will see in the debugger that for received byte XX the value of the timer's CCR1 register is 0000XXXX as the byte is duplicated. We see that the two most significant bytes are zeros but that's because the register is 16-bit only so its value is actually XXXX.

6.14. *Autobauding*

One of the problems with RS-232 is that both devices which use an RS-232 link have to use the same transmission parameters. Because of this fact configuring those parameters is a common element of establishing a serial connection. Fortunately we don't have to set everything manually. While it is not easy to automatically set options such as parity or flow control, speed of the connection can be configured using a simple algorithm. In general it is realized by analyzing one or more bytes transmitted by one of the two devices. If the receiver has prior knowledge about content of those bytes it can determine the speed of transmission and immediately switch to detected speed. If the content is unknown then some guessing may be required.

Here I'd like to present a simple approach. One of the devices sends 80h byte and the other measures the time of low logical state on the RXD line. Having this time it is easy to calculate speed of transmission. Where does the 80h value come from? For easy calculations we would like to have eight 0 bits. The transmission starts with a start bit which, as you know, is always zero. Therefore we need seven more zeroes. Hence the last bit of the transmitted byte must be one. As the bits are transmitted from the least significant one the bit which has to be set to 1 is the most significant one. The 8-bit number which has the most significant bit set is 10000000b = 80h.

6.14.1. AVR code

The code is short and simple:

```
#include <avr/io.h>

void init_Timer() {
    TCCR1B = _BV(CS11)|_BV(CS10); //clk/64
}

void init_UART() {
```

```
      UCSRB = _BV(RXEN)|_BV(TXEN)|_BV(RXCIE); //EX enable, RX
int enable
      UCSRC = _BV(URSEL)|_BV(UCSZ1)|_BV(UCSZ0); //n,8,1
}

void sendByte(uint8_t data) {
      while(!(UCSRA & _BV(UDRE)));
      UDR = data;
}

int main() {
      init_Timer();
      init_UART();
      while(PIND & _BV(PIND0));
      TCNT1 = 0;
      while(!(PIND & _BV(PIND0)));
      uint16_t i = TCNT1;
      i >>= 1;
      i--;
      UBRRL = i;
      UBRRH = i >> 8;
      sendByte('O');
      sendByte('K');
      sendByte('\n');
    while(1);
}
```

To measure time of the logical low state we use 16-bit Timer1. Using the *TCCR1B* register we configure it to work with the clock 64 times slower than system clock. When the timer is configured we initialize USART but don't set speed of transmission. Having everything configured we wait for the RXD line to change its state to 0. The RXD line is pin 0 of port D in ATmega32 so we are monitoring this pin. As a remainder I'd like to point out that we are interested in actual voltage on the pin so we are checking a bit of the *PIND* register, not *PORTD*. From *PORTD* we can read only what we have written there by ourselves and here we are not interested in it. When the start bit comes we reset timer's counter so we can measure the time correctly. When the *RXD* line comes back to 1 we read content of the timer. To obtain a value suitable for USART configuration we have to divide read value by two and decrement by one. Then we write 8 less significant (lower) bits to the *UBRRL* register and 8 more significant (higher) bits to the *UBRRH* register. The higher the speed the lower the value we write. That's why we usually don't use *UBRRH* – the value used is smaller than FFh and the higher half is zero. But if you have high speed clock (e.g. 16 MHz) and low transmission speed (e.g. 300 bps) then you need to set *UBRRH* correctly. When USART is configured we send "OK" string with a new line character for the other device to know that autobauding in done. If the string is received correctly then the device which has sent 80h byte knows that the autobauding finished correctly. In this example there is only one try. If you want you can improve the algorithm. The initiating device can notify the other device that it has correctly received the "OK" string. If there is no such notification then the device which configures itself should wait for another sync byte.

6.14.2.   STM32 code

The code for STM32 is a little more complicated.

```
#include "stm32f10x.h"

void init_USART();
void init_USART_Int();
void init_Timer();
void sendByte(uint8_t c);
void init_GPIO();

void init_GPIO(){

        RCC_APB2PeriphClockCmd(RCC_APB2Periph_GPIOA, ENABLE);

        GPIO_InitTypeDef GPIO_InitStructure;
        GPIO_InitStructure.GPIO_Pin = GPIO_Pin_9;
        GPIO_InitStructure.GPIO_Speed = GPIO_Speed_2MHz;
        GPIO_InitStructure.GPIO_Mode = GPIO_Mode_AF_PP;
        GPIO_Init(GPIOA, &GPIO_InitStructure);

        GPIO_InitStructure.GPIO_Pin = GPIO_Pin_10;
```

```
        GPIO_InitStructure.GPIO_Mode = GPIO_Mode_IN_FLOATING;
        GPIO_Init(GPIOA, &GPIO_InitStructure);
}

void init_USART(){
        RCC_APB2PeriphClockCmd(RCC_APB2Periph_USART1, ENABLE);
        USART_InitTypeDef USART_InitStructure;
        USART_InitStructure.USART_BaudRate = 9600;
        USART_InitStructure.USART_WordLength =
USART_WordLength_8b;
        USART_InitStructure.USART_StopBits = USART_StopBits_1;
        USART_InitStructure.USART_Parity = USART_Parity_No;
        USART_InitStructure.USART_HardwareFlowControl =
USART_HardwareFlowControl_None;
        USART_InitStructure.USART_Mode = USART_Mode_Tx |
USART_Mode_Rx;
        USART_Init(USART1, &USART_InitStructure);
}

void init_USART_Int() {
        NVIC_PriorityGroupConfig(NVIC_PriorityGroup_0);
        NVIC_InitTypeDef NVIC_InitStructure;
        NVIC_InitStructure.NVIC_IRQChannel = USART1_IRQn;
        NVIC_InitStructure.NVIC_IRQChannelSubPriority = 0;
        NVIC_InitStructure.NVIC_IRQChannelCmd = ENABLE;
        NVIC_Init(&NVIC_InitStructure);
        USART_ITConfig(USART1, USART_IT_RXNE, ENABLE);
}

void init_Timer(){

        RCC_APB1PeriphClockCmd(RCC_APB1Periph_TIM2, ENABLE);

        TIM_TimeBaseInitTypeDef TIM_InitStruct;
        TIM_InitStruct.TIM_ClockDivision = TIM_CKD_DIV1;
        TIM_InitStruct.TIM_CounterMode = TIM_CounterMode_Up;
        TIM_InitStruct.TIM_Period = 0xFFFF;
        TIM_InitStruct.TIM_Prescaler = 7;
        TIM_TimeBaseInit(TIM2, &TIM_InitStruct);
        TIM_Cmd(TIM2, ENABLE);
}

void sendByte(uint8_t c) {
```

```
        while(USART_GetFlagStatus(USART1, USART_FLAG_TXE) ==
RESET);
        USART_SendData(USART1, c);
}

void USART1_IRQHandler(void)
{
        if (USART_GetITStatus(USART1, USART_IT_RXNE) != RESET){
            USART_ClearFlag(USART1, USART_IT_RXNE);
}
}

int main() {
        init_USART();
        init_Timer();
        init_GPIO();
        while(GPIO_ReadInputDataBit(GPIOA, GPIO_Pin_10));
        TIM2->CNT = 0;
        while(!GPIO_ReadInputDataBit(GPIOA, GPIO_Pin_10));
        uint16_t i = TIM2->CNT;
        TIM_Cmd(TIM2, DISABLE);
        USART1->BRR = i;
        USART_Cmd(USART1, ENABLE);
        init_USART_Int();
        sendByte('O');
        sendByte('K');
        sendByte('\n');
        while(1);
}
```

First of all we initialize USART and a timer like in AVR example. We also initialized USART pins. Then we wait for start bit. When it is detected we start to count using timer. Its prescaler is set to 7 which means that the clocking frequency is 8 times lower than system clock. When the measurement is finished, we acquire obtained value, disable the timer and use the value to configure USART1's baudrate. In theory we should decrement the value by one but it turns out that the value is a little bit lower than it should be so we don't decrement it further. You can read in the manual that the *BRR* register holds a fixed-point value, consisting of an integer and fractional part. It doesn't change here anything however. When the baudrate is set we enable USART and its *RX* interrupt. In the end we send the "OK" string.

6.14.3.   PC code

The C# code is very simple. It basically sends a byte and waits for a line.

```
using System;
using System.Windows.Forms;
using System.IO.Ports;

namespace SerialSample14
{
    public partial class Form1 : Form
    {
        SerialPort ComPort = new SerialPort("COM1");
        string result = "";
        public Form1()
        {
            InitializeComponent();
            ComPort.BaudRate = 9600;
            ComPort.Parity - Parity.None;
            ComPort.StopBits = StopBits.One;
            ComPort.DataBits = 8;
            ComPort.Handshake = Handshake.None;
            ComPort.DataReceived += OnSerialDataReceived;
            ComPort.Open();
        }

        private void OnSerialDataReceived(object sender,
SerialDataReceivedEventArgs args)
        {
            result = ComPort.ReadLine();
            lblResult.Invoke(new
EventHandler(lblResult_DataReceived));
        }

        private void lblResult_DataReceived(object sender,
EventArgs e)
        {
```

```
        lblResult.Text = result;
    }

    private void btnSync_Click(object sender, EventArgs
e)
    {
        lblResult.Text = "";
        int b = int.Parse(comboBox1.Text);
        ComPort.BaudRate = b;
        ComPort.DiscardInBuffer();
        ComPort.Write(new byte[] {0x80}, 0, 1);
    }
  }
}
```

When the button is pressed, the program takes text from the drop-down list at converts to an integer. This way we get the speed user selected from the list. The value is set at baud rate of the serial port. We clear the serial port's input buffer. Then the 80h byte is sent. When the "OK" string is returned the *ReadLine()* method writes it to the *result* variable. Finally the string is displayed using a label.